GW01246514

The Eye in General Practice

The Eye in General Practice

C. R. S. JACKSON

M.A., D.M.(Oxon.), D.O.M.S., F.R.C.S.Ed.,
Ophthalmic Surgeon, Royal Infirmary, Edinburgh, formerly
Assistant Ophthalmic Surgeon, Glasgow Eye Infirmary,
and Clinical Tutor in Ophthalmology, Oxford University

Seventh Edition

CHURCHILL LIVINGSTONE
EDINBURGH LONDON AND NEW YORK 1975

For B.R.J.

CHURCHILL LIVINGSTONE
Medical Division of Longman Group Limited

Distributed in the United States of America by Longman Inc., New York, and by associated companies, branches and representatives throughout the world.

First edition . . 1957
Second edition . . 1960
Third edition . . 1964
Fourth edition . . 1967
Fifth edition . . 1969
Sixth edition . . 1972
Seventh edition . . 1975

ISBN 0 443 01251 2

Printed in Great Britain by Bell & Bain Ltd., Glasgow

Preface to the Seventh Edition

This edition conforms to the pattern established by its predecessors and the book remains an attempt to describe the commoner diseases of the eye in a way useful to the practitioner and student.

In the present revision, opportunity has been taken to rectify some omissions and to remove the description of rarer entities whose inclusion no longer seems justified.

C. R. S. JACKSON

Edinburgh, 1975

Preface to First Edition

Most practitioners admit to having but slight knowledge of diseases of the eye and this is probably due largely to the difficulties involved in demonstrating ophthalmic conditions to students, to the apparent complexity of the subject, and to its confusing terminology.

The present work is not intended as a complete textbook of ophthalmology suitable for the man who is studying the subject as a speciality. An attempt has been made to describe some of the commoner conditions which are to be found in the eye, and to point to the many ways in which the eye may reveal disease of other systems, or in which diseases of the body as a whole may manifest themselves in the eye.

With the special needs of the general practitioner in mind, the objects of the book are as follows.

1. To describe the common diseases of the eye.
2. To indicate the ways in which dangerous diseases of the eye may be recognised, and to show the reasons for seeking specialist advice. The general lines of treatment are described.

Practioners are often naturally unwilling to burden an already overloaded hospital department with what may be a trivial condition, but all of us who work in Eye Departments know that the examination of the eye is often difficult in the surgery, or in the patient's home, and in the absence of specialised apparatus. (It is sometimes difficult enough even when such aids are available.) It is much better for the practitioner to send his patients for specialist examination than for defective vision to be allowed to develop simply because there were no obvious signs of disease.

Perhaps my own outlook on the problem is influenced by the

fact that I have practised in a single-handed general practice, some distance from a town and from an eye specialist. The difficulties with which general practitioners are faced in connection with eye disease, and the dislike of putting a patient to the expense of a time-consuming, and perhaps unnecessary, journey to hospital were very real to me at that time.

3. The third object is an attempt to help the practitioner in the interpretation of reports which he will receive from the specialist. Patients, when they attend a busy hospital, very often go away with hazy notions of what they have been told, and of the treatment that has been suggested. It is to their doctor that they will go for reassurance and advice. This book is partly an attempt to help the practitioner to discuss with the patient the ins and outs of his condition in the light of the specialist's report.

C. R. S. JACKSON

Edinburgh, 1957

Contents

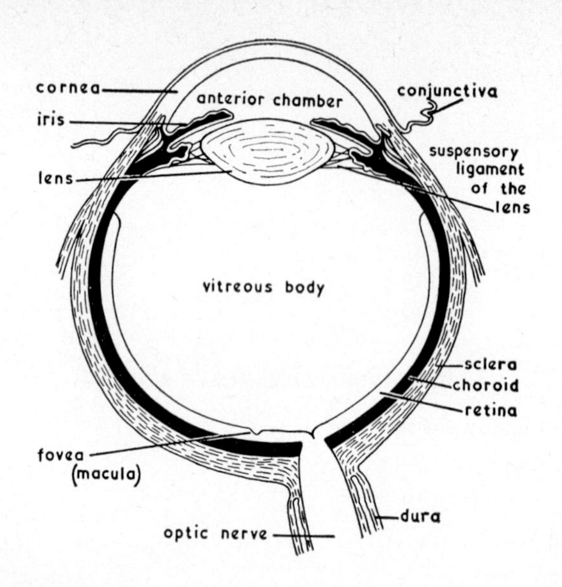

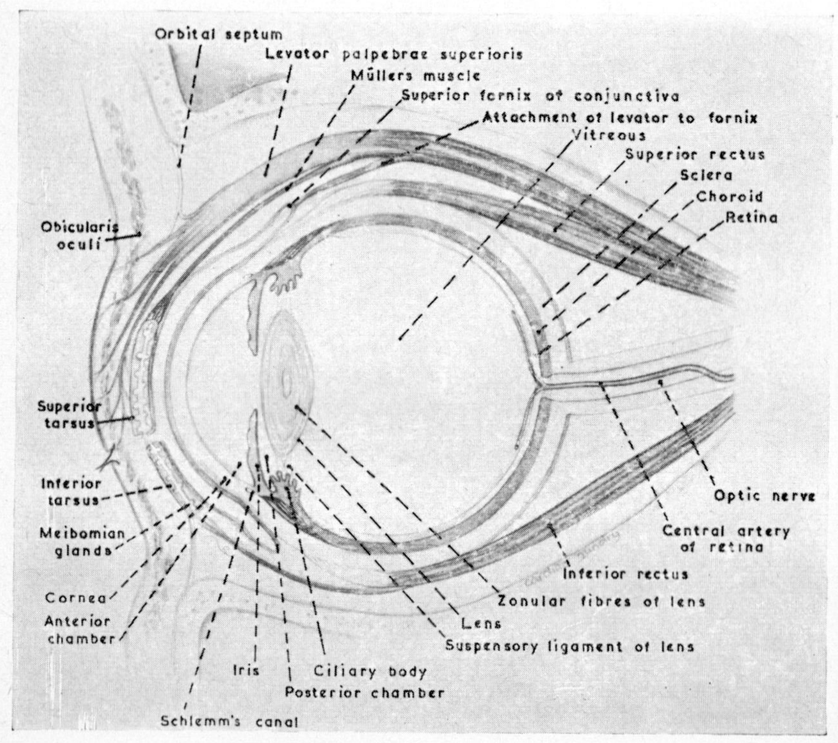

Fig. 1 Anatomy of the Eye.

Part One: Diseases of the Eye

CHAPTER ONE

Examination of the Eye

There are few subjects more difficult to describe than the methods of examining the eye, but there are none in which attention to detail and careful observation are more rewarding. Constant practice and the examination of many eyes lead to greater facility and the readier interpretation of departures from normal.

THE HISTORY

The taking of an accurate and complete history is the mark of a competent physician, for in the history lie the clues to the diagnosis. In the ophthalmic history, attention should be paid to the nature of the visual symptoms, the possibility of an injury, and the occurrence of previous similar trouble. If pain is present its character and distribution are important, as are the results of any measure adopted by the patient in an effort to find relief. For example, the man with a corneal ulcer or foreign body will say that the eye is more comfortable when it is closed and that blinking increases his discomfort. On the other hand, the sufferer from intraocular inflammation has a deep-seated pain, often referred to the cutaneous distribution of the trigeminal nerve; this type of pain, like that of glaucoma, is often worse at night.

The general practitioner will usually bear in mind the possibility of any systemic cause of ocular disease, such as diabetes, rheumatism, or thyrotoxicosis. Nor should the family history be neglected. Myopia, squint, glaucoma, certain tumours, and various degenerative conditions have a strongly hereditary tendency and an enquiry into the occurrence of ocular troubles in the patient's relations will often be of assistance.

For Anatomy of the Eye see Figure 1.

EXAMINATION

The Estimation of Visual Acuity

While one may sometimes be misled by the presence of good visual acuity in an eye suffering from serious ocular disease, it is more likely that neglect of this part of the examination will lead to the false assumption that such disease is not present, simply because no abnormality is to be seen.

It is, therefore, unjustifiable to fail to attempt some assessment of function, though it has been argued that subjective testing of vision is difficult if no formal test-types are available. It is certainly true that no mathematical expression of visual acuity can be obtained in the absence of the correct test objects, but there are very few houses in which there is no clock on the mantelpiece, or a newspaper, which can be used for the purpose. The test should be made while the patient is wearing his spectacles, if any.

If a formal test card of letters is available, this is used at 6 metres distance, or about 20 feet, and evenly illuminated. The letters diminish in size from above downwards. Each row of letters is calibrated, according to the distance at which the normal eye can read it. Thus the usual third row from the top—marked '24'—could normally be read at 24 metres. Working at 6 metres we represent the recorded visual acuity as a fraction, in which the numerator is the distance at which we are working, in this case 6 metres, and the denominator the distance at which the normal eye can read the line in question. If therefore the eye being tested can, at a distance of 6 metres, only read the line normally read at 24 metres, we say that the visual acuity is 6/24. A scale can thus be constructed, ranging from a normal acuity of 6/6, through 6/9, 6/12, 6/18, 6/24 and 6/36 to 6/60. In the event of no letter on the card being read at 6 metres, the patient is moved closer to the card, when the vision may be recorded as 3/60 or even 1/60, depending on the distance at which the top letter may be made out. If the top letter cannot be read at any distance we test if the eye in question can count the examiner's fingers when held up in front of it. If so, this vision is 'Counts fingers' (C.F). Failing this, we find out if the eye can see a hand moving in front of it (H.M.). If not, we test for the ability to perceive light (P.L.). If perception of light is absent, the eye is totally blind and has 'No P.L.'.

In the case of children, the above test using formal letters may not be possible, and some type of illiterate test is called for. Of these tests the most useful is the 'E' test. Capital E's in diminishing size and turned into various directions are shown to the child, who responds by indicating the direction in which the 'legs' of the E's are pointing. Many children aged between two and three years

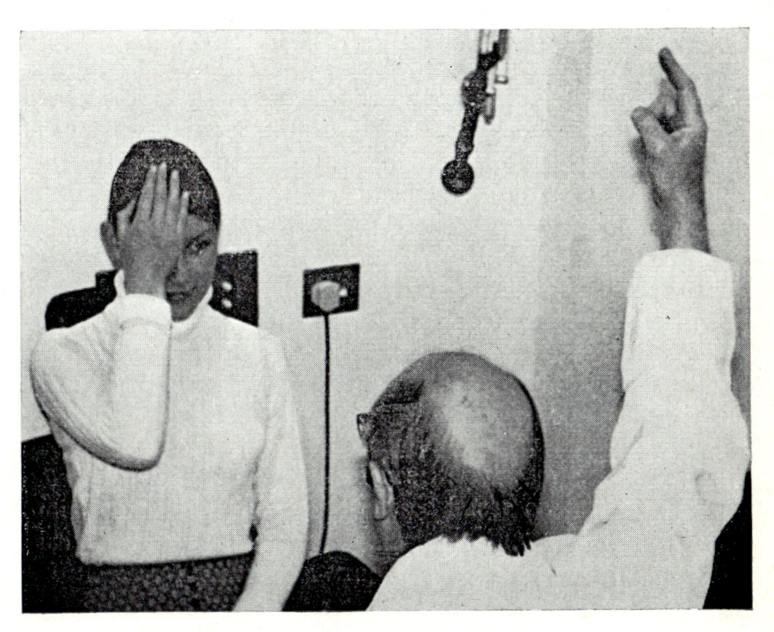

Fig. 2 Testing Field of Vision.

can readily learn this test and give accurate results. A test which is often more useful in the pre-school child is the Sheridan-Gardiner test. The child identifies a letter by matching it against one of a number of such letters on a card in his hand.

One other subjective test which is sometimes applicable is the examination of the field of vision. This can be done by the confrontation method (Fig. 2).

The patient and examiner sit facing one another about an arm's length apart. While the patient has one eye closed, the examiner moves a test object (his finger, or the head of a hat-pin) from the periphery inwards, midway between their two heads. By doing this in the various meridians and ensuring that the patient maintains steady fixation on the examiner's face, the practitioner can compare the extent of the patient's visual field with his own and will be able to detect the presence of field defects, such as hemianopia. It is important to check the periphery of the visual field and not to confine this test to the central area.

Ocular Motility

Under this heading must be included the movements of the lids, the presence or absence of proptosis, and the ability of both eyes to follow a test object smoothly and synchronously as it is moved to the various positions of the gaze. By testing the movements of the lids, one has a clue to the integrity of the third and seventh cranial nerves, also the sympathetic. The presence of lid-lag or lid retraction may raise a suspicion of thyroid disease, and limitation of movement of an eye in one or more directions may point to a lesion of the third, fourth or sixth nerves. Diplopia, in one or more directions, may be elicited.

Before detailed examination of the lids and globe is begun, a test of tactile sensation over the cutaneous distribution of the fifth nerve may be indicated.

The Lids

Having tested the lid movements, and having excluded gross deformities, the relation of the lid to the globe is examined, as both lids should lie smoothly in contact with the eyeball in all positions (see entropion and ectropion). The lid margins may have to be examined under magnification, to exclude the presence of deformed or misplaced eyelashes, which are a potent cause of ocular discomfort. At the same time the presence and normal position of the lacrimal puncta are confirmed, and pressure over the lacrimal sac at the nasal side of the eye will reveal any mucocele formation. Peri-orbital tenderness can be excluded at the same time. Finally, both lids should be everted in order that their under aspect can be examined.

While this is being done, any blepharospasm or photophobia is noted.

The Conjunctiva and Cornea

The normal eye is almost white. A few conjunctival vessels are to be seen, particularly in the fornices. Oedema (*chemosis*) of the conjunctiva will be noted, as will vascular engorgement and dilatation. The distribution of these vascular signs and the type of vessel involved (whether superficial or deep) is often of assistance in deciding the site of any inflammatory process in the eye. Foreign bodies and conjunctival swellings are usually seen without difficulty.

In the cornea, its bright anterior surface and normally complete translucency are the two most important points to note. It is a fair general rule that a cornea which is bright and shiny, and which reflects light clearly and without distortion from all parts of its convex anterior surface, is a healthy cornea. In fact, any corneal injury (with one exception) and any corneal ulcer (again with one exception) will leave a permanent opacity in the corneal stroma and, very often, a depression in the epithelium which distorts the reflection of rays from that part of the surface. A further feature of corneal inflammation is a tendency to vascularisation, and the character and distribution, at least of superficial vessels, can usually be made out without elaborate apparatus. Cellular deposits on the posterior corneal surface which, as will be seen, are a feature of some types of intraocular inflammation, can often be made out with the simple *corneal loupe*, or magnifier (×8), and a *focusing torch*. These two instruments are the only ones required for an adequate examination of the anterior parts of the eyeball as far back as, and including, the iris and the anterior surface of the lens. The torch should preferably be of the pen-type and should have a switch, rather than pressure control. It can then be held between finger and thumb while the other fingers of the same hand can be rested on the patient's face or used to steady the lids (Fig. 3). The hand supporting the corneal loupe rests on the patient's forehead and allows accurate focusing, while a finger restrains blinking movements of the upper lid. It is essential for the observer's eye to be very close to the lens, to achieve the largest field of view.

The Anterior Chamber

This is the space between the cornea and the iris. It is full of aqueous humour and its depth is uniform (about 3 mm.). Variations in depth are sometimes present, and it is important to look for abnormal constituents of the fluid, blood, pus and foreign bodies, which sink into the lowest part.

Experience allows one to judge the depth of the anterior chamber. In some eyes, particularly those which are appreciably hypermetropic, and those in which the lens is tending to become swollen with age, dilatation of the pupil with drugs has to be undertaken with care owing to the risk of precipitating an attack of closed-angle glaucoma.

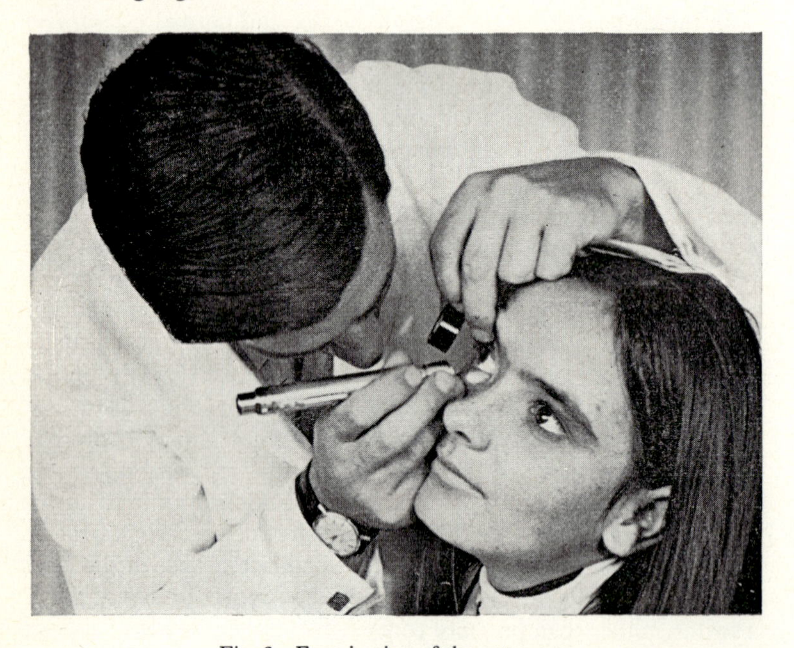

Fig. 3 Examination of the outer eye.

The Iris

The colour of the two irides should be compared and any abnormality of iris pattern noted, for it is the general rule that the pigmentation of the iris is the same on the two sides in each individual, and any difference may indicate injury or disease.

For the examination of pupil function, good daylight is most suitable, as it is necessary that the two eyes are equally illuminated. While the patient looks into the distance the eyes are both shaded and are then uncovered in turn. In this way the *direct reaction* to light is elicited and also the *consensual response* of the shaded eye to the light falling on its fellow. The reaction of accommodation is tested by asking the patient to look first into the distance and then at an object held about six inches from the face. This should lead to a brisk contraction of both pupils.

The pupils are normally round and are equal in size. Any size difference, or departure from circular, may be evidence of local disease (inflammation or trauma), or of interference with the nervous pathways.

The Lens

It is not possible to examine the transparency of the lens by direct light. If a beam of light from a torch is thrown on the

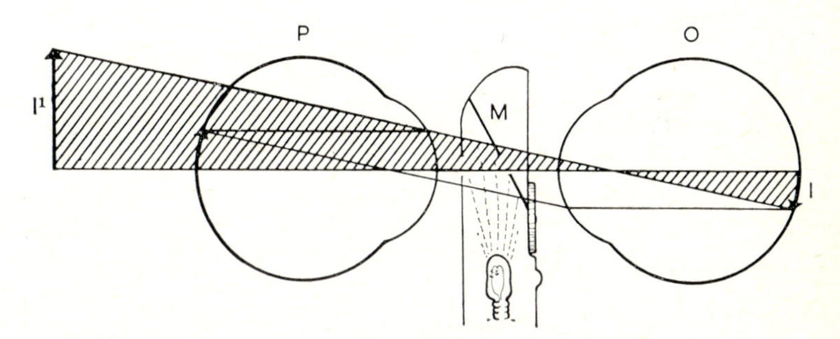

Fig. 4 Ophthalmoscopy.

An area of the patient's eye (P) is illuminated by light reflected from the perforated mirror (M) of the ophthalmoscope. The image of this lighted area is formed on the retina of the observer's eye (O). This image (I) is projected outwards to I^1, at a distance of approximately 25 cm, the usual distance for near vision. In this case the magnification is of the order of $\times 15$, and depends on the relationships of two 'similar triangles', which are shown shaded in the diagram.

eye, the pupil appears black in youth, but an opalescence gradually develops in the lens as age increases and an apparent haze may be seen by this method of examination giving rise to the impression that lens opacities are present.

The only way in which the transparency of the lens can be estimated is by transmitted light, the transparent ocular media being seen against the red reflection of the light from the retina and choroid. This method of examination demands that the observer's eye be placed in the same line as that of the incident light, and is directly comparable with the way in which animals' eyes appear luminous when seen from behind the headlights of a car at night (Fig. 4).

It is, therefore, necessary that some form of *ophthalmoscope* be available for the production of the required optical conditions, and some guidance may be desirable as to the best types of instrument. Since the introduction of the self-luminous electric ophthalmoscope the bulk of the technical difficulties in the use of these instruments have been removed. The cheaper models depend on the use of a prism to deviate the beam of light and these, of which the May ophthalmoscope is the best known example, are adequate for the examination of the normal eye, especially with the pupil artificially dilated. They have the disadvantage that the amount of light available is small and that the prism tends to become scratched or displaced in use.

The difficulty with regard to providing oneself with an ophthalmoscope is primarily that of expense, but in general it can be said that the more expensive instruments are the more efficient. They have greater light concentrating power and a more refined optical system. It is these advantages which allow the use of the instrument under conditions in which small models are of little assistance.

A good ophthalmoscope will last for most of the practitioner's working life, and the advice that one gives to intending purchasers is to buy the best instrument that they can afford after talking to the owners of various types of ophthalmoscopes and having the opportunity of trying the various models.

Difficulties in the Use of the Ophthalmoscope

In the use of the ophthalmoscope the beginner is faced with a number of difficulties, some of which are avoidable. The main troubles which confront the beginner with the ophthalmoscope are:

1. An inadequate instrument.

2. A small pupil, which makes detailed examination of the eye very difficult, particularly in the elderly. Dilatation of the pupil with 1 per cent homatropine is justifiable if it is considered necessary to obtain a view of the deeper parts of the eye. As an alternative cyclopentolate, 1 per cent (Mydrilate, Cyclogyl), is useful. Atropine should be avoided, as its effects are unnecessarily powerful and prolonged. After the examination one drop of pilocarpine 2 per cent is instilled into the eye to counteract the effect of the mydriatic. (Mydriasis = pupillary dilatation. Miosis = pupillary constriction).

In rare cases, if the anterior chamber is shallow, dilatation of the pupil restricts the drainage of aqueous humour and leads to elevation of the intraocular pressure. This will show itself by aching pain in and around the eye, some disturbance of vision and haziness of the cornea. Onset of symptoms may be delayed for several hours. Repeated application of pilocarpine is indicated, and, possibly, further treatment for congestive glaucoma (q.v.).

3. The presence of extraneous light in the room. General medical wards are one of the most difficult places in which to examine patients' eyes. It is often necessary to wait until evening before attempting examination of the fundus. In an emergency both the patient and the examiner may be draped in a light-proof sheet to enable the examination to be carried out in relative darkness.

4. Improper handling of the instrument. The rules with regard to the use of the ophthalmoscope are: patient's right eye, your right eye, your right hand; patient's left eye, your left eye, your left hand. The patient's right eye is examined from the right side and the left eye from the left side.

5. Myopia. The optical conditions in myopia are such that the retinal image is enlarged, and this is one of the conditions which gives rise to most trouble. If the myopia is of high degree, it is sometimes a help to examine the patient while he wears his own spectacle correction. To do this will reduce the apparent size of the retinal image and enable a more easy examination.

6. Inadequate appreciation of normal appearances. If the many possible variations of the normal are to be appreciated, it is essential that every available opportunity be taken of examining eyes. It is only in this way that the practitioner gets used to using his ophthalmoscope under varying conditions and learns to sum up the various types of normal eye: for example, the variations in size and shape of the physiological cup at the optic disc, the changes in colour of the normal fundus, and such 'normal abnormalities' as opaque nerve fibres and colloid bodies in the choroid.

The technicalities of ophthalmoscopy could not very well be described shortly in print. Sufficient to say that with no lenses in the sight-hole of the ophthalmoscope, and the instrument held at a distance of some six inches away from the patient's eye, a red reflex will be apparent in the pupil. This reflex will either be undisturbed or it will be broken up by opacities in the ocular media, whether these be in the cornea, the lens or vitreous. Approaching the eye more closely and focusing the instrument with the range of lenses available by rotating the control wheel, it is possible to see fundus details (Fig. 5). The first thing will probably be a retinal vessel, often a vein, as these are more easily visible than the arteries. Following the vein towards the centre

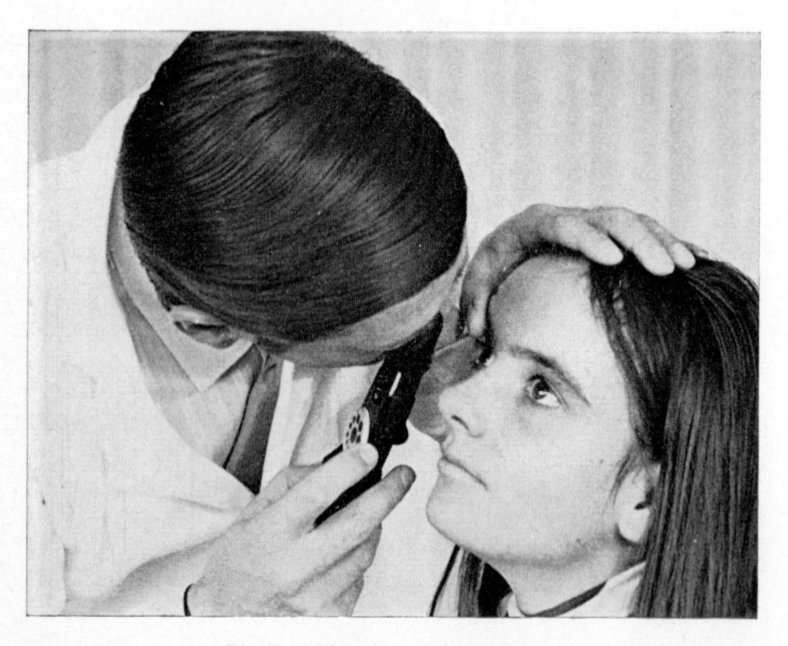

Fig. 5 Using the ophthalmoscope.

of the globe, the optic disc will be seen lying to the nasal side of the posterior pole and having a paler colour than the surrounding fundus (Fig. 6). The colour of the optic disc varies greatly from one subject to the next as does its shape and the size of the central depression or the physiological pit. It is from this pit that the retinal vessels emerge, and they break up into upper and lower temporal and nasal branches spreading out towards the periphery. It will be seen that the retinal vessels emerge from the nasal side of the optic disc and that those which are directed temporally take a wide sweep above and below to reach the temporal retina. Almost all the vessels are branches of the main trunk, but there are often small arteries, two of which can be seen in Figure 6, which arise from the ciliary circulation and supply part of the retina to the temporal side of the optic disc. These are cilio-retinal arteries. Arteries and veins cross one another, but one never sees an artery

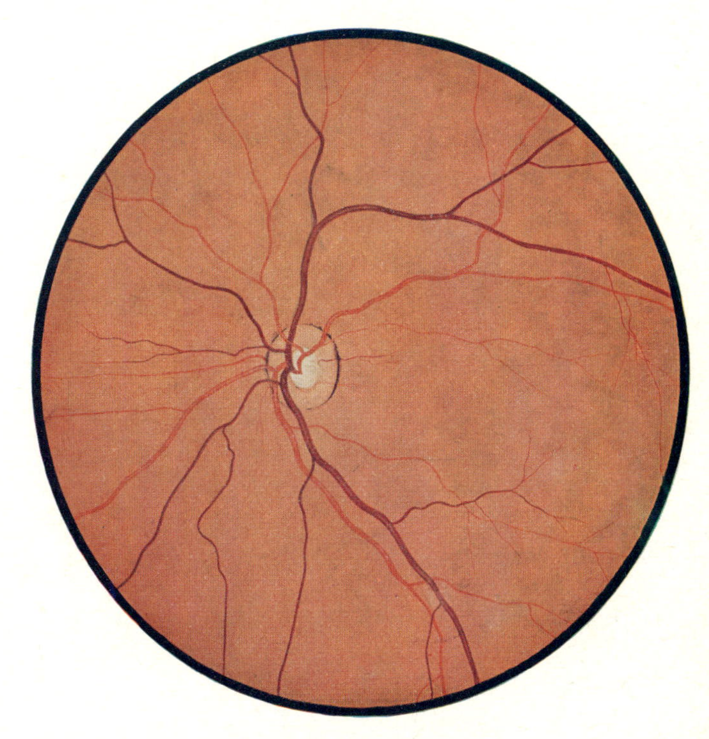

Fig. 6 Normal fundus. Left Eye.

crossing another artery, or a vein crossing a vein. To the temporal side of the optic disc and approximately one and a half disc diameters away from it, there is an avascular area which contains the region of most distinct vision, the macula lutea. This region shows a bright reddish reflex and the reflex is more readily visible in youth. As a guide to size, we can say that the normal optic disc is about 1.5 mm in diameter.

The colour of the general fundus depends for the most part upon the quantity and distribution of pigment in the underlying choroid. It varies from an extremely dark colour seen in heavily pigmented individuals and Negroes to a pale pink which is common in fair-haired individuals and is seen to its most fully developed extent in albinism.

It is only by repeated examination of many eyes that an assessment of the range of normality can be achieved, and it is only by experience that one can appreciate the finer degrees of abnormality.

It is not proposed to discuss the many other and more specialised examinations of the eye, as these are available only in special departments. They include perimetry, tonometry, examination with the biomicroscope, estimation of colour vision and many other tests, both subjective and objective.

CHAPTER TWO

The Eyelids

ANATOMY

The skin of the eyelid is thin and there is no subcutaneous fat, a feature which accounts for the ready swelling in any inflammatory lesion or of any systemic cause for oedema. In their normal position, with the eyes looking straight forward, the margin of the upper lid crosses the upper third of the cornea, while the lower lid lies at the inferior limbus (junction of cornea and sclera). Exposure of the sclera below the cornea is one of the first signs of proptosis. The posterior concavity of the lids, which conforms to the curve of the cornea, is maintained by the presence of the two *tarsal plates*, condensations of connective tissue which give the lids their stiffness. It is in the substance of these plates that the *tarsal or meibomian glands* lie. These glands discharge by a number of ducts opening at the lid margin. The tarsal plates are continuous with the *medial and lateral palpebral ligaments*, by which the lids are anchored to the bone at each end.

At the anterior part of their free margins, the lids carry two or three rows of eyelashes directed away from the globe. As in the case of hair follicles elsewhere, the lashes are associated with small sebaceous and sweat glands. The inner end of each lid is free of lashes and carries, at its posterior border, a small *papilla* on the apex of which opens the *lacrimal punctum*, the upper end of the lacrimal drainage apparatus. The normal lacrimal punctum is directed slightly backwards toward the globe, and cannot be seen unless the lid is everted.

Movements of the lids are carried out by two voluntary and one involuntary muscles. The *orbicularis oculi* is innervated by the facial nerve and forms a circle of fibres around the orbit. It is responsible for blinking movements and for the forcible protective closure occurring in injury or inflammation. Elevation of the lid is performed by the *levator palpebrae superioris* supplied by the third cranial nerve, and also by a small sympathetically innervated muscle, *Müller's muscle*, overaction of which causes lid retraction in thyrotoxicosis.

For Anatomy of the Eye see Figure 1.

WOUNDS OF THE LIDS

Wounds of the lids demand special attention, not only on account of the dangers resulting from improper handling, but also because they may accompany a severe injury to the eyeball itself. The oedema which occurs produces great swelling of the tissues and examination of the globe may be difficult. It should not be neglected on this account. If the injury is seen early, before much swelling of the lids has developed, then it should be possible gently to open the eye and examine the cornea. The pupil reaction can be assessed and the normal circular position of the pupil noted. In a conscious patient a rough test of vision should be carried out. The presence of free blood in the conjunctival sac makes one suspicious of the presence of ocular injury.

Lacerations of the lids bleed considerably, but this bleeding can almost always be controlled by pressure. It is on account of the very considerable blood supply that these injuries heal very well and every piece of tissue can, therefore, be preserved during the repair. Excision of wound edges or of badly torn tissue is neither necessary nor desirable.

Horizontal cuts often need no repair, as they gape little, but vertical lacerations must be sutured. Special care is required if the lid edge is involved, as deformity here will not only lead to the formation of an unsightly notch or coloboma, but will also cause distortion of the lid margin and interference with the normal position of the lashes. In injuries close to the medial canthus a fine probe should be passed into the lacrimal passage. If the lacrimal canaliculus is found to be divided immediate repair holds out the only hope of cure, and this repair should be the responsibility of an ophthalmic surgeon.

A danger which is present in destructive injuries of the lids is exposure of the cornea. If immediate repair is not to be undertaken, it is essential that the cornea be prevented from damage, either by the use of plenty of ointment and a pad, or of two or three temporary stitches to hold the torn lids in a position to cover the globe.

If repair of a full-thickness lid wound has to be attempted, the lid should be repaired in two layers. Sutures from the conjunctival aspect of the lid will include the conjunctiva and tarsal plate, and should be of fine plain catgut, while the skin stitches will include the skin and orbicularis muscle. Normally, these sutures can be removed after five days.

INFLAMMATION OF THE LIDS

BLEPHARITIS. Inflammation of the lid margin occurs in two forms: *squamous* and *ulcerative*. Of these, the former is by far the more common.

SQUAMOUS BLEPHARITIS. The patient complains of chronically irritable eyes and a degree of conjunctivitis is constantly present. At the lid margins fine crusts are to be seen at the roots of the lashes and removal of these reveals an intact skin surface. A very frequent association is the presence of seborrhoea capitis, and there is no doubt that this is the predisposing factor. The condition is common in childhood, particularly in fair-skinned individuals, and most children seem to throw off the condition as they grow older.

Treatment is long and tedious. The use of a good soapless shampoo to the scalp, together with some scalp lotion, is often effective in controlling the dandruff. Locally, removal of the crusts from the roots of the lashes and the firm application of antibiotic ointment at night is usually all that is required. Steroids also are useful. They may either be used in the form of drops, such as betamethasone three times a day, or of ointment which is applied to the lid margin at night.

ULCERATIVE BLEPHARITIS. This condition is due to pyogenic organisms and occurs in ill-nourished individuals, particularly children. The crusting at the lid margins is more extensive and is associated with ulceration of the lid. This is accompanied by falling of the lashes and, later, by deformity of the lash follicles and distortion of the growth of the hairs, leading to abrasion of the cornea and ulceration.

Local treatment consists of frequent removal of the crusts and the application of antiseptics. Good results may be obtained with sulphacetamide ointment or with chloromycetin drops by day and ointment at night. Systematic treatment is essential. A full diet with vitamin supplements, a change of air, and a course of ultra-violet light may all be employed.

HORDEOLUM. This, the common stye, is an abscess occurring in one of the glands related to a lash follicle. It is distinguished, therefore, by the fact that pointing occurs in the line of the lashes. Styes tend to occur in crops and are then a source of great irritation, both to the patient and the physician, as treatment is unrewarding. The single stye can occasionally be aborted by pulling out the affected lash. Sulphacetamide or chloromycetin ointment should be used to prevent infection of other follicles while frequent hot bathings relieve the pain and assist resolution. Repeated acute styes are probably best treated by a short intensive course of systemic penicillin; autogenous vaccines have also been advocated. Massive doses of vitamin B complex are reputed to be effective in preventing recurrences.

MEIBOMIAN ABSCESS. This is an infection occurring in a chalazion (q.v.) and is distinguished from a stye by the fact that it points at a distance from the lid margin. It may require surgical drainage.

SWELLINGS OF THE LIDS

Benign

1. CHALAZION. Although this is referred to as a meibomian cyst, it is in fact a granuloma occurring in a meibomian gland (Fig. 7). It gives rise to a chronic swelling in the substance of the lid and sometimes reaches a size of 10 mm in diameter. There is no pain unless abscess formation takes place. The swelling is usually situated at some distance from the lid margin. Infection is commonly followed by the formation of a breach on the conjunctival surface of the tumour and the production of granulation tissue which may be mistaken for a papilloma. Meibomian cysts never resolve spontaneously and incision and curettage is required for their removal. This can readily be carried out under local

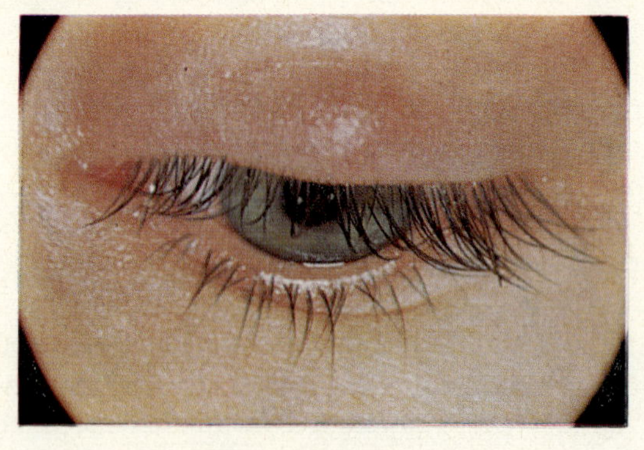

Fig. 7 Chalazion.

anaesthesia, a small incision being made over the cyst on the conjunctival surface and at right angles to the lid margin.

The conjunctiva is anaesthetised with amethocaine, and the affected lid is infiltrated with local anaesthetic through a fine needle. A special meibomian clamp holds the everted lid and assists in haemostasis and the incision is made with a fine scalpel. The contents of the cyst are removed with a small curette. Bleeding ceases after a few minutes of firm pressure over the closed lids and a dressing is usually not required.

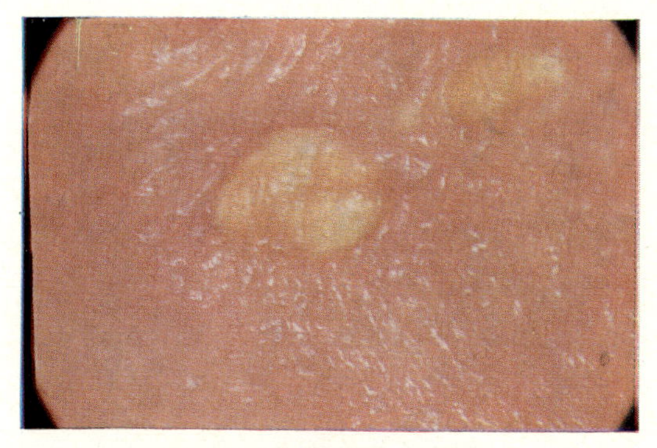

Fig. 8 Xanthelasma.

2. OTHER LID CYSTS. The less common varieties of cyst occurring in the lids include cysts of the glands related to the lash follicles and giving rise to rather watery-looking swellings close to the lid margin. They are easily drained through small incisions after the injection of a little anaesthetic. Dermoid cysts are not uncommon and are found most frequently close to the orbital margin at the angles, where they are often attached to the bone.

3. XANTHELASMA. Elderly people not infrequently show flat yellowish plaques in the skin of the lids (Fig. 8). This is a degenerative change of no pathological significance, but excision or treatment by photocoagulation is sometimes demanded for cosmetic reasons. Recurrence is not uncommon.

4. MOLLUSCUM CONTAGIOSUM. Infection with the molluscum virus gives rise to multiple slightly raised umbilicated swellings close to the lid margin and these may be either single or multiple. They are slow-growing and require to be excised or curetted.

5. ANGIOMA. Spider naevi occur as a congenital condition and often become less obvious as the child grows. Treatment is therefore not required. Cavernous angiomata may be rather more of a problem for they may be of considerable size, increasing when the child strains or cries. Angiomata present as purplish swellings often with dilated skin vessels overlying. The child's parents are anxious about the condition, but active treatment should be postponed as long as possible for a degree of spontaneous regression is the rule and the swelling may become unnoticeable. If treatment is necessary, the choice lies between surgery and radiotherapy. Of these, radiotherapy is preferable, as these tumours are very freely vascularised and often poorly encapsulated, making excision difficult. The implantation of radium needles causes a reaction of fibrosis and consequent shrinkage of the tumour.

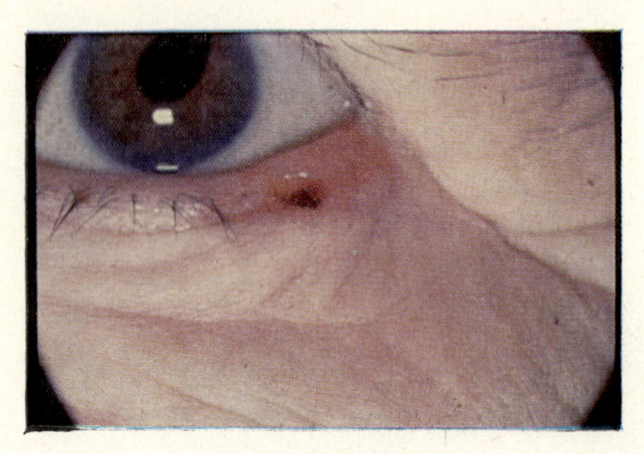

Fig. 9 Rodent ulcer.

MALIGNANT

1. RODENT ULCER. Not only is the basal-celled carcinoma the commonest malignant tumour of the eyelids, but the majority of these growths occurring in the body as a whole are seen on the face (Fig. 9). They are very slow-growing and are only locally invasive. The early lesion is slightly raised and possesses a characteristically rolled edge. Crusting frequently takes place, with apparent healing in the intervals. If untreated, gradual destruction of tissue results, with loss of substance of the lids, invasion of bone and, occasionally, loss of the eye.

The choice of treatment is between surgery and radiotherapy, and opinion is divided as to their relative value. Radiotherapy often gives an excellent result with minimal scarring; though it is not always suitable for lesions at the inner canthus owing to possible interference with tear drainage. Alternatively, complete excision must be undertaken. If the lid edge is involved in the lesion the defect resulting from excision may demand an extensive plastic repair.

2. EPITHELIOMA. This is a less common lesion and often metastasises early to the gland fields. As the tumour is not so radio-sensitive as is the rodent ulcer, wide excision is the best treatment, preceded by diagnostic biopsy. In the absence of adequate early treatment, the prognosis is not good.

MALPOSITIONS OF THE LIDS

ENTROPION

When the margin of the eyelid is turned backwards, the lashes, instead of being directed away from the globe, are turned inwards and irritate the eye. This maldirection of the lashes (*trichiasis*) also occurs in many conditions in which the lid margin is deformed, such as the late results of wounds, chemical burns and ulcerative blepharitis. The rubbing lashes give rise to conjunctivitis and, frequently, to corneal abrasion and ulceration. The resulting opacities may interfere with vision. Inward rotation of the lid edge is most often due to spasm of the orbicularis oculi and may be precipitated by the bandaging of an eye, particularly in an elderly person after an operation, though it does occur spontaneously (senile spastic entropion) (Fig. 10). In a well-developed case the lashes may be actually invisible unless the lid is everted. It is always the lower lid that is involved. The patient complains of a sensation as of something in the eye and, even if the lid is in normal position on first examination, the entropion can often be produced by asking the patient forcibly to close the eye. If the condition causing the spasm is a transient one, such as an attack of conjunctivitis, the entropion will often disappear when the cause of the irritation is removed, or some strapping can be applied to the cheek to hold the lid in place. Failing this, a small plastic operation to the lid is desirable. This is often done as an outpatient procedure, under local anaesthesia.

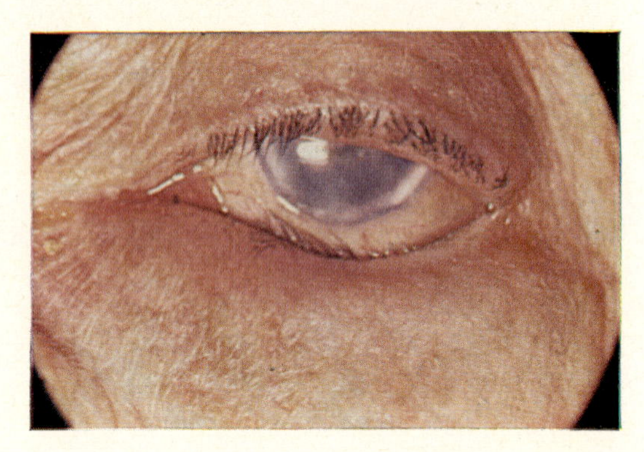

Fig. 10 Entropion.

Ectropion

As in the case of entropion, failure of the lower lid to maintain its normal apposition to the globe may be the result of inflammatory lesions or of wounds. It may also result from loss of tone in orbicularis muscle, as occurs in age (Fig. 11). There may be simply a little falling away of the inner end of the lid from the globe, or the whole lid may be everted and the conjuctiva exposed along its entire length. The result of any movement of this kind is that the tears fail to gain access to the drainage apparatus, and the first

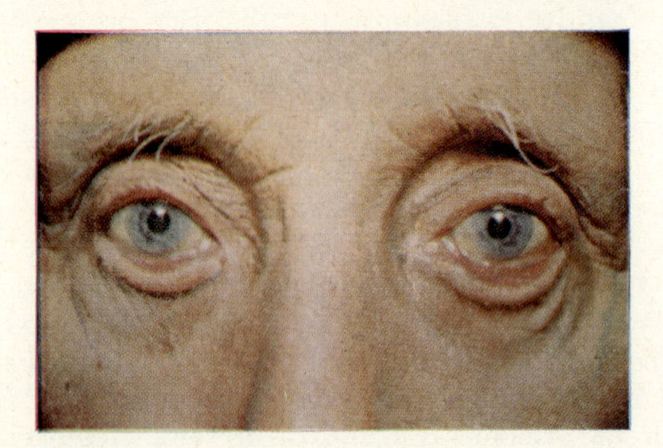

Fig. 11 Ectropion.

symptom, therefore, is epiphora. This is always accompanied by conjunctivitis, which in turn, leads to increased lacrimation. The constant escape of tears on to the face leads to excoriation of the skin and the resultant contracture increases the deformity. Treatment is surgical, and a variety of plastic procedures have been designed to draw the lid back into its normal position. In early cases, where the inner end of the lid alone is misplaced, improvement may be obtained by simply placing some cautery punctures on the conjunctival surface of the lid below the lacrimal punctum. Contraction of the scar tissue so formed will restore the lid to its normal position.

Ptosis

Drooping of the upper lid occurs in paralysis or weakness of the levator of the lid as a part of a third nerve lesion, and as such may be seen at any age and may be due to a multitude of causes, such as vascular lesions, head injuries, intracranial tumours and so on. It is not uncommonly seen as a manifestation of some generalised neurological disorder, particularly myasthenia gravis, and is a feature of tabes. The diagnostic feature of ptosis in myasthenia is that it tends to be worse at the end of the day and may be provoked during examination by making the patient gaze steadily at the examiner's finger held in front of the patient's face and above the horizontal plane.

Of greater importance from the ophthalmic point of view is congenital ptosis, which may be unilateral or bilateral, complete or partial, and which may be associated with a weakness of elevation of the affected eye, due to paralysis of the superior rectus muscle. It is impossible to mistake a child suffering from bilateral congenital ptosis for he has a characteristic 'head-back' attitude in an attempt to see through his reduced palpebral aperture.

Treatment is a matter of urgency only if the cornea is completely covered by the upper lid and the eye is, therefore, not being used and is likely to become amblyopic. This is more likely to obtain in a unilateral case. In other cases, where there is full vision in each eye, correction is best left until the child is seven years old or so, when any operation is easier on account of the greater size of the parts and a degree of spontaneous improvement may well occur before this time. This is particularly true with regard to the *epicanthus* which often goes with congenital ptosis (Fig. 12). As the child's nasal bones develop, so the epicanthic fold is drawn forward and becomes less prominent while a reduction in the amount of ptosis also takes place.

Treatment, if necessary, is surgical, and consists of an operation to shorten the levator palpebrae superioris. The cosmetic results are very good. Another possibility, to be considered in the case of

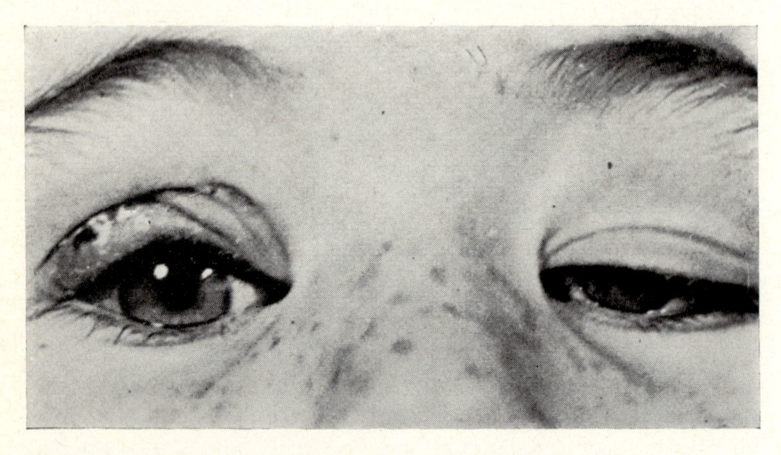

Fig. 12 Epicanthus and ptosis. (Right post-operative.)

an elderly patient or in one in whom the ptosis is due to some active lesion or neurological disease, is the provision of 'crutch glasses'. These consist of a spectacle frame to the upper part of which is fastened a horizontal bar projecting backwards. By its pressure on the upper part of the lid, this bar raises the lid margin and reduces the ptosis. Such spectacles, however, are often uncomfortable in wear, and difficult to keep in adjustment. Most patients do not appreciate them.

CHAPTER THREE

The Conjunctiva

ANATOMY

The conjunctiva is a mucous membrane which lines the lids and is reflected on to the front surface of the eyeball. It is loosely attached to the globe, except at the corneoscleral junction, or limbus, where the conjunctival epithelium becomes continuous with the epithelium of the cornea. The membrane is normally moistened by the secretion of its many glands, and additional lubrication is supplied, under conditions of irritation or inflammation, by the lacrimal gland, which drains, by a dozen or so ducts, into the conjunctival sac at its upper outer corner. This conjunctival sac, however, is a closed space only when the eyelids are closed. It normally is exposed to the air and is, therefore, never strictly sterile, but the growth of organisms is hindered by thc bacteriostatic activity of an enzyme, lysozyme, contained in the tears.

INJURIES TO THE CONJUNCTIVA

TRAUMATIC

1. Lacerations

Lacerations of the conjunctiva occur as a result of contusions to the globe, or of penetration by sharp objects. Both varieties of injury deserve to be treated with a degree of respect greater than their apparent simplicity requires, for they may conceal a penetration of the eyeball itself. For this reason an attempt must be made to assess the visual acuity of the injured eye, and to examine the deeper parts of the eye with the ophthalmoscope. Neither of these measures are likely to be easy in the case of a child, or in the presence of associated corneal damage, but they should not be omitted on this account.

For Anatomy of the Eye see Figure 1.

The conjunctiva heals rapidly if the edges of the wound are in apposition, but gaping wounds must be repaired on account of the tendency for granulation tissue to develop between their lips.

For the repair of such a wound the eye is anaesthetised with two or three drops of amethocaine 1 per cent at intervals of five minutes. A drop or two of adrenaline 1:1000 is of great value in giving a bloodless field.

If no speculum is available an assistant may help to support the lids, while the edges of the wound are lifted with forceps, and the underlying sclera inspected and palpated with the tip of a probe.

No penetration of the eye being present, the wound is repaired with interrupted stitches of fine plain catgut. Some antiseptic ointment is instilled, and the eye is covered for twenty-four hours. After this no dressing is usually needed.

If any doubt as to the integrity of the eye exists, the case should be referred for specialist opinion.

2. Conjunctival Foreign Bodies

When a foreign body is blown into the eye, or a loose lash falls into the conjunctival sac, it is commonly to be found in the lower fornix from which it can easily be removed while the patient looks upwards. If the particle lies beneath the upper lid, not only does it scratch the cornea with each act of blinking, but it cannot be removed without everting the lid to expose its under surface. This manoeuvre can most easily be accomplished by standing behind the patient who is seated. His head can then rest against the operator's chest. The steps in the operation are as follows (Fig. 13):

(*a*) The patient looks downwards with both the eyes open and his chin well up.

(*b*) The operator grasps the eyelashes between finger and thumb of one hand while, with the other, he makes counter pressure with the tip of a pencil or glass rod on the skin of the lid just above the tarsal plate.

(*c*) While the lower part of the lid is thus drawn away from the globe, a movement of the two hands in opposite directions rotates the tarsal plate about its long axis and exposes the conjunctival surface of the lid.

(*d*) If the patient is constantly instructed to continue looking downwards, the lid can be maintained in this position by the pressure of a finger at its lower border.

(*e*) The foreign body is removed with the tip of a probe or the pulp of a finger. Subsequently, the cornea is stained with fluorescein to exclude corneal abrasion and some antibiotic ointment is instilled.

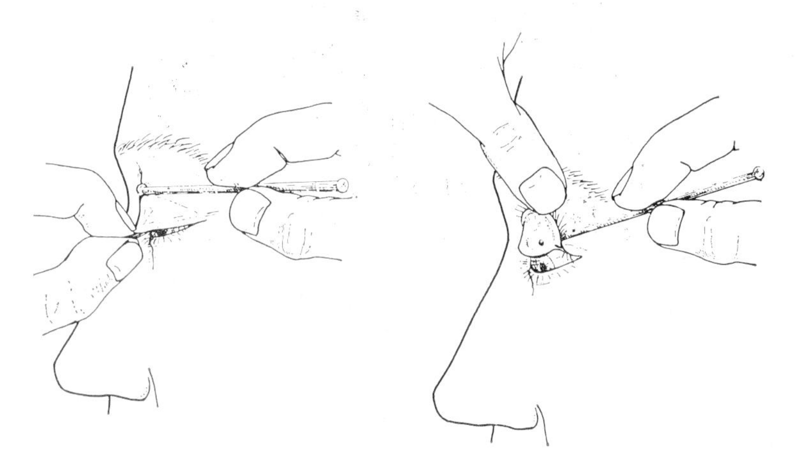

Fig. 13 Eversion of the upper lid.

The upper and lower recesses of the conjunctival sac are unexpectedly capacious, and quite large foreign bodies may be concealed therein. It is not at all unusual for such things as air-gun pellets to be recovered from these spaces.

3. Subconjunctival Haemorrhage

A localised haemorrhage may follow contusions to the eye and often is part of an orbital haematoma (Fig. 14). The haemorrhage is absorbed in the course of a week or two, but, if it is severe it should raise the suspicion of the possible presence of more serious injury, such as fracture of the orbital walls. It should also be remembered that orbital haemorrhage may occur as a result of a remote injury to the skull and follow subarachnoid bleeding.

Occasionally, these haemorrhages occur spontaneously without apparent cause though they may be a manifestation of generalised arteriosclerosis. Spontaneous subconjunctival haemorrhage is often recurrent and occurs mostly in the elderly.

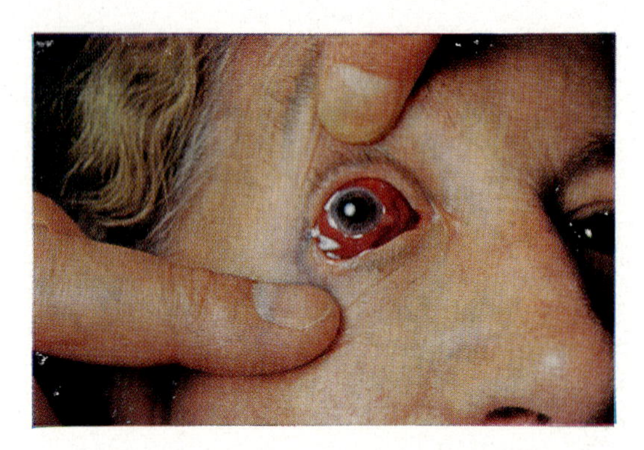

Fig. 14 Subconjunctival haemorrhage.

CHEMICAL

This is a very important group of cases, for energetic first-aid measures may exert great influence on the ultimate result.

Workers in chemical industries or laboratories sometimes receive splashes of liquid acids or alkalis in the eyes. In both cases the resultant reflex flow of tears dilutes the irritant at once, but immediate irrigation of the eyes with an abundance of water may be a sight-saving measure. The injured man's companions should hold his head under a tap and assist him to open his eyes while the water runs over them. The subsequent management of the case depends on the severity of the burning, but, in general, alkalis are more dangerous than are acids, and some of the worst cases of corneal opacity occur after contact with liquid ammonia.

Such injuries are rather specialised, but one chemical which requires special mention is *lime*. Solid lime may be splashed in the face in a variety of occupations, especially among builders and plasterers, often through the powder flying up during careless mixing of cement.

There is immediate lacrimation and blepharospasm, making examination difficult.

In the surgery, treatment of the case is as follows:

1. Anaesthetise the eye with amethocaine.

2. Remove all solid particles of lime from the conjunctiva, using forceps or cotton wool swabs.

3. Irrigate the eye thoroughly with water or, preferably, with a saturated solution of glucose, which converts any residual lime into insoluble calcium gluconate.

4. Stain cornea with fluorescein.

5. If the cornea is clear and there is no evidence of conjunctival necrosis in the fornices, instil antibiotic ointment.

6. If necrosis is present, or the cornea is hazy, refer the case for specialist opinion and treatment.

CONJUNCTIVITIS

Inflammation of the conjunctiva can occur in any degree of severity, from a devastating purulent infection, to a mild, low grade irritation, often of long duration and with minimal objective signs.

Again, while most of the more acute varieties are frankly infective in aetiology, the causation of many types is often a matter of conjecture, and treatment is, therefore, arbitrary and symptomatic.

GENERAL SYMPTOMATOLOGY

Conjunctivitis is the commonest cause of the 'red eye' and this redness is associated with a sensation of grittiness or, as the patient often says, of 'sand in the eyes'. Inflammation is most marked on the inner aspects of the lids, in the fornices, and towards the angles, a feature which is important in the distinction between this disease and many conditions affecting the cornea and deeper parts of the eye.

In addition, irritation of the mucous membrane leads to increased discharge, manifest by a sticking together of the eyelids in sleep, due to drying of the secretion around the eyelashes. In the absence of this symptom of stickiness of the eyes in the mornings, a diagnosis of conjunctivitis is in doubt.

It is to be remembered also, that simple conjunctivitis is almost invariably (or becomes) bilateral. If conjunctivitis remains unilateral, search should be made for a source of irritation (foreign body, dacryocystitis, inturned lashes, etc.).

PRINCIPLES IN TREATMENT OF CONJUNCTIVITIS

It is important to avoid covering the eye. If occlusion is needed for the sake of appearance, a pair of dark glasses may be used, as these allow ample ventilation. The popular celluloid eyeshade is to be forbidden in any infection of the outer eye. It is impervious to moisture, allows no ventilation and converts the infected conjunctival sac into a moist, warm culture chamber. These eyeshades are impossible to sterilise, and are often passed on from one member of a family to another in the treatment of any sore eye.

The question of eye bathing is also relevant. Many people bathe their eyes regularly each night, in a routine similar to that in which they wash their teeth. This practice has little or nothing to be said in its favour. The eyes are naturally protected by the movements of the lids and by the presence of the lacrimal secretion, which contains a bacteriostatic enzyme, lysozyme. To bathe the eyes dilutes this tear fluid. In addition, some of the domestic remedies used for eye bathing are potent ocular irritants, and may themselves lead to a conjunctivitis (see Conjunctivitis due to drugs). The only firm indication for the use of eyebaths is for the removal of discharge, and for this nothing is more suitable (or more readily available) than warm normal saline. Certain astringent lotions may sometimes be ordered for the treatment of specific conditions, but their use is limited.

BACTERIAL INFECTIONS

1. Muco-purulent Conjunctivitis (Fig. 15). (Catarrhal Conjunctivitis, pink-eye)

This is the commonest acute variety of conjunctivitis and often runs through boarding-schools with great rapidity. The aetiology is varied, but the offending organism is usually a staphylococcus or a virus. There is increased lacrimation and flakes of muco-pus, floating in the tears in the lower fornix, are to be seen when the lower lid is pulled down.

The condition usually responds rapidly to treatment with sulphonamides or antibiotics. These drugs are applied in the form of drops during the day, while ointment at night helps to keep up the concentration and it prevents the annoying sticking together of the eyelids in the mornings.

Ideally, all cases would be treated under bacteriological control. In practice, most cases can be treated satisfactorily with sulphacetamide (drops 10 per cent, ointment 6 per cent). As an alternative there is chloramphenicol (drops 1 per cent, and ointment 1 per cent).

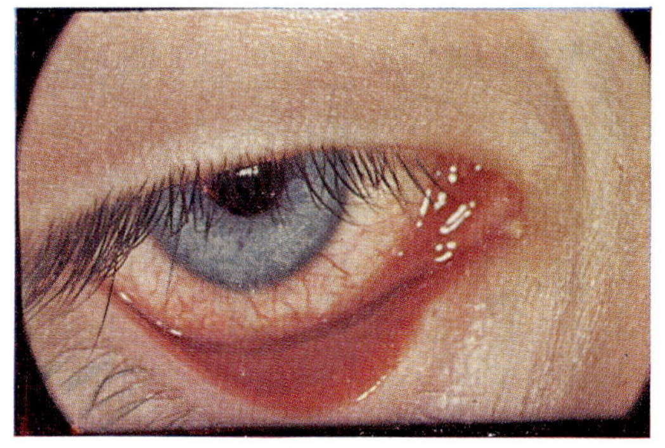

Fig. 15 Conjunctivitis.

In the early stages of treatment drops should be applied frequently (two-hourly at least) and the condition reviewed at the end of forty-eight hours. Absence of response at the end of this time implies resistance of the organism and the need for a change of treatment.

Many cases of infective conjunctivitis show a good response to treatment at first, and then settle down to a condition of mild irritation of the eye, showing no further change. The procedure now adopted is to withhold all treatment for a few days (except perhaps saline bathing). There is no doubt that conjunctival irritation may result from over-treatment.

If several days have elapsed between the onset of the condition and its first coming under treatment, the conjunctiva in the fornices is considerably engorged and thickened, with a surface which has lost its normal lustre. A valuable step in treatment is to paint the conjunctiva with silver nitrate 1 per cent, an operation readily done in the surgery. The eye being anaesthetised with amethocaine, the inner aspect of the upper and lower lids is thoroughly swabbed with silver solution, using cotton wool wrapped round a probe or match-stick. This treatment can be repeated at intervals of three or four days if a response is being obtained, but silver nitrate should not be given to patients to use at home. Its prolonged use may result in unsightly permanent brown staining of the conjunctiva, especially in the lower fornix (*argyrosis*).

2. Angular Conjunctivitis

Here, the chief redness of the eye is at the angles and there is often some dermatitis and fissuring of the skin close to the outer canthus. Zinc sulphate $\frac{1}{4}$ per cent drops used three times a day clear the condition by interfering with the metabolism of the infecting organism. Zinc oxide cream is used for the skin lesions.

3. Chronic Conjunctivitis

This is a disease, in the aetiology of which low-grade infection plays a part, but many cases of conjunctival irritation will be found to be associated with some other local or general condition. The list of possible causes is very long, and includes dacryocystitis, malformations of the lid (especially senile ectropion), ingrowing eyelashes, conjunctival foreign bodies, infection in nose or throat, exposure to irritating dusts, fumes and smoke, and errors of ocular muscle balance. This list is by no means comprehensive, but it shows that a search should be made for any possible local or systemic source of irritation. Errors of refraction do not figure in the above list. The reason for this is that there has been, in the past, too much emphasis laid on the part played by such errors in the aetiology of ocular irritation, and that their importance is doubtful.

Treatment is often unrewarding. If no local cause of irritation is found, investigation of the patient's general health and the nature of his work may give a clue. Any infection present in the conjunctival sac or tear duct should be treated in the ways mentioned. Discomfort is often relieved by the use of zinc sulphate drops.

4. Purulent Conjunctivitis

Until a few years ago this was one of the chief blinding diseases, particularly among children. It is now relatively uncommon and always curable if treated early and energetically. The danger of the condition lies in the fact that it may be complicated by corneal ulceration and subsequent opacification.

(*a*) IN INFANTS (*Ophthalmia Neonatorum*). This is a notifiable disease and the official definition includes any case of purulent discharge from the eyes of a child occurring within twenty-one days of birth.

Infection occurs during passage through an infected birth canal and usually shows itself within two or three days as gross oedema of the lids, between which pus escapes when an attempt is made to open the eye. The conjunctiva is seen to be congested and swollen (*chemosis*).

Treatment is best carried out in an institution equipped for this work and requires the undivided attention of at least one nurse during the early stages. The pus is removed by gentle irrigation as it forms and antibiotic drops are instilled every few minutes at first, and then at lengthening intervals. The infant is also given a course of systemic antibiotic. Improvement is seen within a few hours, and the case is usually cured in two or three days. While most cases of ophthalmia neonatorum were in the past due to the gonococcus, only 25 per cent are now due to this organism; others are caused by staphylococci or pneumococci.

(*b*) IN ADULTS. Purulent conjunctivitis in adults is often gonorrhoeal and involves the right eye in most cases, due to auto-inoculation.

5. Other Bacterial Infections

This group includes tuberculosis and syphilis, both of which are rare causes of conjunctival lesions.

VIRUS CONJUNCTIVITIS

Trachoma

Chronic conjunctivitis and keratitis due to trachoma comprise one of the chief blinding diseases of the world. The disease is endemic in India, Africa, Russia and the Eastern Mediterranean area. Only sporadic, imported cases occur in the United Kingdom. Treatment until recently was unsatisfactory, though some of the newer antibiotics seem of value.

Some cases present with an acute onset, but the majority begin as a chronic conjunctival irritation, with photophobia. Over the course of months, scarring and deformity of the upper lid take place and there develops an associated keratitis, with vascularisation and opacification of the cornea, especially in its upper part.

There is no doubt that the devastation produced by this disease is being reduced by an intensive programme of education and routine antibiotic treatment in the schools, but it remains a serious economic and human problem.

ALLERGIC CONJUNCTIVITIS

Spring (Vernal) Catarrh

A seasonal irritation of the eyes, occurring mostly in boys and characterised by intense itching, lacrimation and the development of flat papilliform thickenings of the conjunctiva, mostly on the inner aspect of the upper lid (Fig. 16). (There is a rarer bulbar form in which the disease occurs on the globe.) Eosinophils are found in the tears.

The disease is usually self-limiting, not only as regards the individual attacks, but also the recurrences. These tend to be less and less severe, and eventually cease.

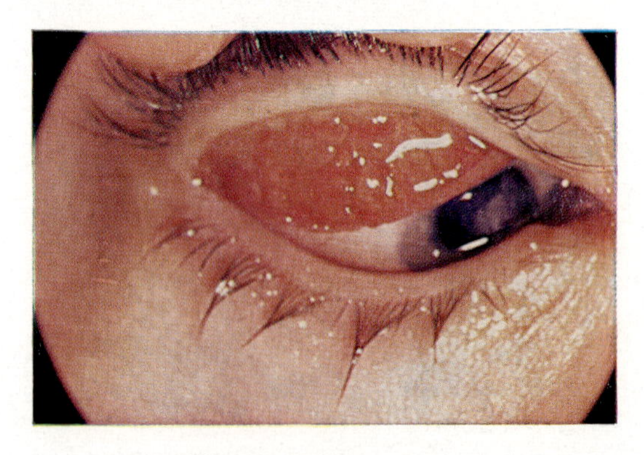

Fig. 16 Spring catarrh.

Treatment is largely symptomatic, but the availability of hydrocortisone has enabled us to control the disease in the majority of cases. Hydrocortisone drops are used two to three times daily, depending upon the severity of the symptoms, and they produce a rapid increase in the patient's comfort. Skin testing for allergies and subsequent desensitisation may be helpful. In severe cases superficial X-ray treatment to the conjunctival surface of the upper lid has been used.

Phlyctenular Disease

Occurring in undernourished children from overcrowded homes, this condition almost certainly has some relationship to tuberculosis. There may be a family history of tuberculosis or the patient (often a child) may be a Mantoux-positive. While it is primarily a conjunctival disease, comprising single or multiple milky nodules close to the limbus, phlyctenular conjunctivitis shows a tendency to invade the cornea. On the cornea the lesions are round or tongue-like vascularised ulcers, at first close to the limbus.

Clinically, the child has recurrent attacks of red, sore eyes, with severe photophobia if the cornea is involved. This photophobia is the most distressing and characteristic feature of the picture.

In treatment, improvement in the child's general state is often curative in itself. It is remarkable how a case of phlyctenular disease will heal within a few days if the child is admitted to hospital, sometimes only to relapse again on return to the home surroundings. Plenty of fresh air, a full diet, and vitamin supplements are invaluable. Atropine 1 per cent b.d. and chloramphenicol ointment t.i.d. may be needed if the eyes are very irritable.

Follicular Conjunctivitis

This is another disease of youth and one which usually does not give rise to much disability. There may be mild redness and irritability of the eyes.

Examination shows granularity of the conjunctiva, particularly in the fornices. This is due to enlargement of the lymphoid follicles in the submucosa, a condition which is part of the generalised hypertrophy of lymphoid tissue in youth.

Some bland ointment to be used at night is the only treatment required, as the disease is self-limiting.

Conjunctivitis due to Local Sensitisation

A number of substances applied locally are liable to cause conjunctivitis of varying degrees of severity, often with associated dermatitis of the skin around the orbit (Fig. 17). As in the case of other allergic states the patient complains of intense itching and there is much lacrimation. In some cases, particularly those following the use of drugs, there may be a generalised sensitisation dermatitis with an erythematous eruption on the body.

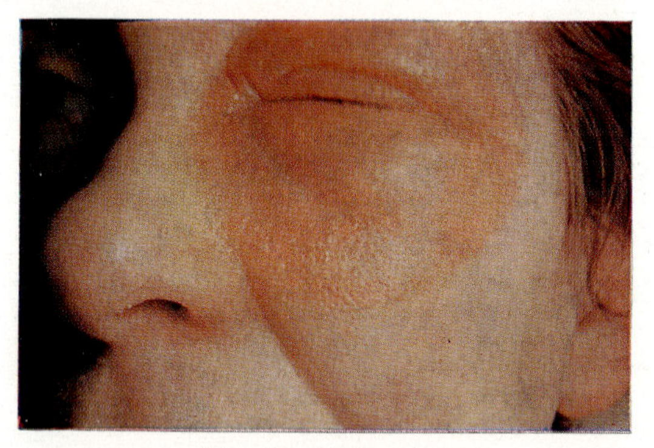

Fig. 17 Drug rash.

Among the commoner substances causing the condition are:

1. Drugs—atropine, cocaine, eserine, boracic, mercury, sulphacetamide and pilocarpine.
2. Cosmetics—eyelash pencil and face powder.
3. Vegetable substances—Primula.

DEGENERATIVE CONDITIONS

Pinguecula

This is a common condition in the elderly. Close to the medial and lateral sides of the cornea an area of subconjunctival degeneration occurs, leading to the development of creamy triangular plaques with their bases towards the limbus. No symptoms are caused and no treatment is indicated, though excision is sometimes requested for cosmetic reasons.

PTERYGIUM

This is an active condition, the aetiology of which is not understood, but in which heat and dust probably play a part, as it is common in hot, sandy countries.

An area of conjunctival degeneration occurs at the limbus in the horizontal meridian and the apex of the cone is directed towards the centre of the cornea (Fig. 18). The infiltration, which is

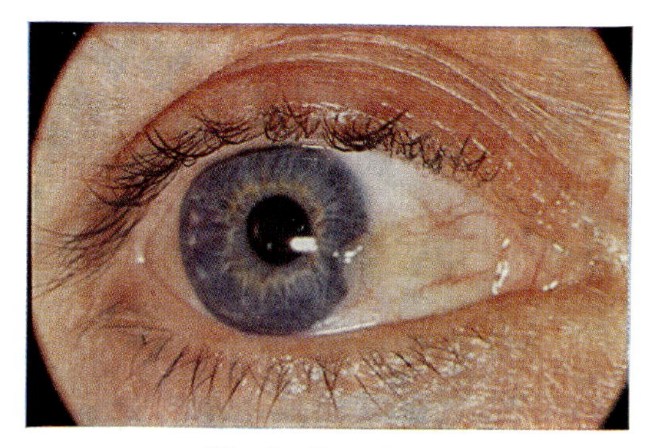

Fig. 18 Pterygium.

preceded by a zone of corneal opacity, gradually advances on to the cornea, drawing with it a tongue of vascularisation from the conjunctiva. If the pupillary area is threatened, excision or transplantation of the affected area of conjunctiva is indicated.

CONJUNCTIVAL TUMOURS

These are rare and include the following:

BENIGN

NAEVUS

Often close to the limbus, a conjunctival mole may be of any colour from coffee to jet black. It is slightly raised and has a nodular surface. Some of these moles become more prominent and more heavily pigmented at puberty. Unless excision is demanded for cosmetic reasons, no treatment is indicated. As in the case of melanomata elsewhere, however, any increase in the size or vascularity of the lesion should be regarded with suspicion. Malignant change occurs in a small proportion of cases and is an indication for wide excision and, probably, clearance of the orbit.

PAPILLOMA

Local excision presents no difficulty.

MALIGNANT

CARCINOMA

Very rare.

MALIGNANT MELANOMA

See above.

CHAPTER FOUR

The Watery Eye

ANATOMY AND PHYSIOLOGY

The lacrimal gland lies in the upper outer corner of the orbit, and its secretion drains into the upper conjunctival fornix. From here the tears pass by the movements of the eyelids towards the inner corner of the conjunctival sac. On each lid, close to its inner end, there is a minute *lacrimal punctum*, invisible unless the eyelid be everted, for in its normal position the lacrimal punctum points slightly backwards towards the globe. From the puncta two *lacrimal canaliculi* run medially towards the bony medial wall of the orbit, where they enter the *lacrimal sac*, opening into the *naso-lacrimal duct* which, in its turn, drains into the inferior meatus of the nose below the inferior concha. In addition to the lacrimal gland itself, the conjunctiva is plentifully supplied with mucous glands and with small accessory lacrimal glands. It is probable that the secretion of these additional glands maintains the moistness of the conjunctiva under normal conditions, and that active secretion of the lacrimal gland itself is only called for under conditions of irritation and emotion.

The tears enter the lacrimal passages partly by capillary attraction which draws them into the narrow canaliculus, and partly by the muscular action of the lids compressing and dilating the lacrimal sac.

Abnormal watering of the eyes may be brought about by either over-production of tears or by obstruction to the outflow.

EPIPHORA DUE TO OVER-PRODUCTION OF TEARS

1. Irritation of the conjunctiva or cornea. This may result from a conjunctival or corneal foreign body, a corneal ulcer or simple conjunctivitis, or deeper ocular disease.

For Anatomy of the Eye see Figure 1.

2. Reflex epiphora from irritation of the fifth nerve, either on the face or in the nose. This type of epiphora also occurs, particularly in sensitive individuals, on exposure to bright light. The mechanism of this effect is not clear.

3. Emotion.

The treatment of any of these types of watery eye is the treatment of the cause.

EPIPHORA DUE TO OBSTRUCTION TO THE OUTFLOW OF TEARS

It is convenient to discuss epiphora of this type according to the anatomical parts involved.

Interference with the Punctum

There may be congenital absence or misplacement of the lacrimal punctum, it may be everted and thus fail to pick up the tears from the conjunctival sac, or there may be stenosis or obstruction following inflammation or wounds. In treating this kind of condition, it is necessary to restore the lacrimal punctum to its normal position, most often by employing a small plastic operation, and to ensure its patency, perhaps by repeated probing and dilatation.

Occasionally a loose eyelash gets washed into the punctum, producing epiphora and a characteristic patch of redness where it rubs against the conjunctiva. The appearance is of an inturned eyelash, but the relationship of the condition to the lacrimal punctum is diagnostic. Removal of the foreign body produces a gratifyingly rapid cure.

Obstruction in the Canaliculus

The commonest type of interference with the canaliculus is its involvement in wounds and scars. If a wound of the canaliculus is treated in an early stage, it is sometimes possible to restore its continuity by the use of a fine plastic cannula over which the cut ends of the canaliculus are repaired. More often, however, the canaliculus becomes irretrievably involved in the scar tissue and it is only at a late date that attempted reconstruction of the lacrimal drainage apparatus can be undertaken. The results of this type of surgery are disappointing.

There is a fungus, a streptothrix, which sometimes grows in the lacrimal canaliculus and may become calcified. The patient complains of chronic irritation of the eye with watering and the offending mass of organism can usually be massaged out of the canaliculus through the dilated punctum.

Obstruction of the Lacrimal Duct

1. IN INFANTS. Infantile obstruction of the naso-lacrimal duct is a common condition and is the commonest cause of unilateral conjunctivitis in the baby.

The naso-lacrimal apparatus arises developmentally from a solid cord of ectodermal cells which become folded into the face along the groove between the fronto-nasal and maxillary processes. Subsequent canalisation of this cord of cells leads to the development of the lacrimal sac and the naso-lacrimal duct. In some cases canalisation is incomplete, and there remains a membrane occluding the lower end of the duct. In yet other cases the obstruction is not a physical blockage, but is due to the accumulation of cast-off epithelial cells and debris, preventing the flow of tears down to the nose.

There is usually no abnormality of the eye during the first few weeks of life, and then it is noticed that one eye is more sticky than the other and tends to water. Pressure with the finger over the lacrimal sac will often produce a reflux of mucopurulent material due to dilatation of the lacrimal sac.

The fact that some of these obstructions are due merely to accumulation of debris means that conservative treatment is worth a trial over a period of some weeks. This treatment consists of repeated expression of the contents of the sac by firm finger pressure at the inner corner of the eye. The mother should be instructed to do this many times a day. If antiseptic drops are used at the same time, a cure will result in a proportion of cases, though it may be some weeks before improvement is seen. If conservative treatment fails to produce a cure the duct will require to be probed in order to overcome the obstruction. This probing requires to be done under a general anaesthetic in an Eye Department.

2. IN THE ADULT. Chronic obstruction of the naso-lacrimal duct is often due to chronic dacryocystitis, a disease which is far more

common in women than in men and is most often seen after the menopause. It must be remembered, however, that the lacrimal apparatus may occasionally be involved in pathological processes arising in neighbouring structures, particularly the antrum, the nose and other nasal sinuses, or in fractures of the bones of the face. The lacrimal sac itself may also be the site of primary neoplasms.

Dacryocystitis in the adult may be either acute or chronic, and acute dacryocystitis most commonly is superimposed upon the chronic condition.

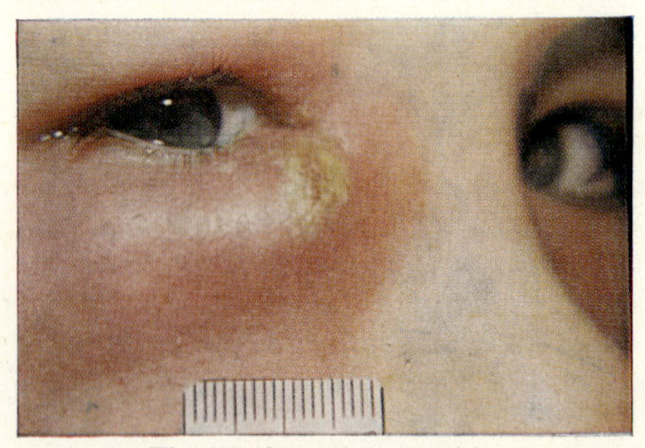

Fig. 19 Acute dacryocystitis.

Acute dacryocystitis is characterised by pain and swelling at the inner corner of the eye and often by systemic upset. If the swelling becomes localised it will be seen that it is centred below the medial palpebral ligament, a fact which distinguishes this type of inflammatory swelling from cellulitis spreading forwards from ethmoid or frontal sinus (Fig. 19). If the inflammation does not resolve as a result of treatment with systemic antibiotics, incision is called for, since spontaneous rupture of the lacrimal abscess may lead to the development of a lacrimal fistula which discharges tears on to the cheek.

Chronic dacryocystitis leads to almost constant epiphora and conjunctivitis, and it may or may not be associated with the development of a lacrimal mucocele from which mucopus can be expressed by finger pressure at the inner corner of the eye.

Although chronic dacryocystitis may seem to be a relatively trivial condition, apart from the watering of the eye to which it gives rise it is a potential source of danger. Not only is the obstructed lacrimal sac liable to the development of an acute attack of inflammation: it is also a source from which inflammatory material is constantly entering the conjunctival sac. This means that any injury to the eye carries a potential risk of infection with resultant corneal ulcer, and any operation upon the eye is fraught with danger unless this chronic source of sepsis can be removed.

Treatment by probing of the naso-lacrimal duct in adults is not to be recommended, for the duct is narrow and the presence of chronic inflammatory tissue leads to a rapid scarring of the passage created. In most patients the treatment of choice involves the creation of a new channel to the nose. This operation, which can be carried out under local anaesthesia, and which is known as dacryocystorhinostomy, consists of the anastomosis through the medial wall of the orbit of the lacrimal and nasal mucous membranes. The tears, therefore, are enabled to enter the nose without passing down the obstructed nasolacrimal duct.

CHAPTER FIVE

The Cornea

ANATOMY

The cornea comprises the transparent anterior segment of the outer coat of the eyeball. Its diameter is about 12 mm, though it is not quite round when seen from the front. In thickness the cornea is about 1 mm thick at the periphery and slightly less at the centre. Despite its relative thinness, however, the cornea is remarkably tough and the risk of opening the eye inadvertently during the removal of a foreign body is negligible. The bulk of the structure is composed of flattened *corneal lamellae* laid down mostly parallel to the surface. Interspersed among the lamellae are a number of rounder cells, the *corneal corpuscles*. The exposed surface of the cornea is covered with epithelium continuous with that of the conjunctiva. Immediately beneath the epithelium are a large number of *free nerve endings* whose presence accounts for the extreme sensitivity of the cornea to touch and for the pain resulting from corneal injury or inflammation.

The main bulk of the cornea is bounded by two structureless membranes which play a part in the maintenance of corneal transparency. These are, in front, *Bowman's membrane* and, behind, *Descemet's membrane*.

At the periphery, where the cornea blends with the sclera, it takes part in the formation of a specialised region where most of the drainage of the intraocular fluid take place. The inner aspect of the cornea is here less compact and it allows access of the aqueous humour to a circular channel, *the Canal of Schlemm*, which is in communication with the veins on the surface of the eyeball.

Although the transition from cornea to sclera at the corneo-scleral junction (the limbus) is apparently an abrupt one, histologically the fibres of the cornea run uninterruptedly into the sclera at this point and the transition is, therefore, due to a difference of physical behaviour rather than of structure.

For Anatomy of the Eye see Figure 1.

CONGENITAL AND HEREDITARY DISEASE

Keratoconus (Conical Cornea)

In certain eyes the cornea, after having been normal in the early years of life, gradually loses its normal curvature and becomes increasingly prominent, with resultant irregular refraction of the entering light. Visual acuity falls, usually during early adult life, and cannot be adequately improved by spectacles. There is no known medical means of arresting the process, but the vision can be greatly improved by the provision of contact lenses, which replace, by a smooth curve, the irregularly refracting surface of the cornea. The operation of corneal grafting is being increasingly employed in the relief of this condition, if useful acuity cannot be achieved with contact lenses.

Corneal Dystrophies

These comprise an obscure group of conditions, the aetiology of which is unknown. Occurring more commonly in adult life and with a familial tendency, various types of corneal opacities develop and interfere increasingly with vision. There is a complete absence of inflammatory signs and treatment is of no value. Corneal grafting holds out hope of visual improvement.

Of particular importance in the elderly is Fuch's endothelial and epithelial dystrophy. This is a degenerative central corneal change leading to opacity. The condition sometimes responds to corneal grafting, which should precede the removal of a cataract if this is present.

Congenital Corneal Dermoid

This congenital tumour occurs at the limbus and covers a variable extent of the cornea (Fig. 20). It is raised above the surface, shiny and pearly-white in colour, and sometimes carries hairs on its surface. It is non-progressive and interferes with vision only if it covers the pupil area. Excision is usually required on cosmetic grounds, though a permanent defect remains in the shape of an opaque scar at the site of the tumour.

TRAUMATIC CONDITIONS

Corneal Abrasion

This is an injury of the cornea which goes no deeper than the epithelium. The possible causes are innumerable, but among them are the flying foreign body, the corner of a sheet of paper, a twig or plant stalk, and a baby's finger nail. There is intense pain and lacrimation, a sensation as if there were something in the eye, and the eye is seen to be injected with blepharospasm and photophobia. The area of epithelial damage can be demarcated by staining with fluorescein. A sterile paper strip impregnated with

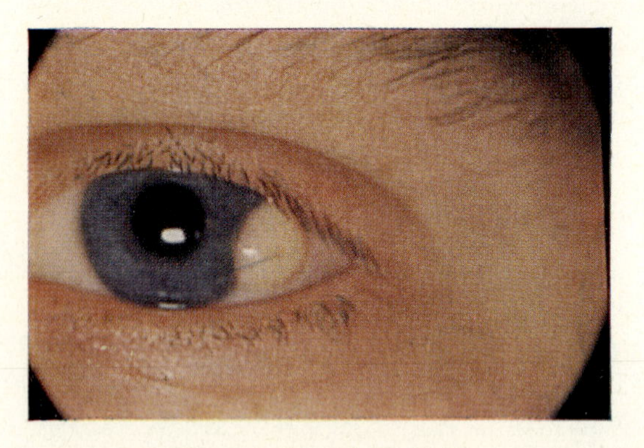

Fig. 20 Corneal dermoid.

the dye is moistened with saline or with the patient's own tears and is allowed to remain momentarily in the lower conjunctival fossa. Blinking distributes the stain over the eye, and any breach in the corneal epithelium shows up as a bright green staining area (Fig. 21). Examination is easier if the eye is anaesthetised with a drop of local anaesthetic. Having excluded the presence of a foreign body either on the cornea or in the conjunctival sac, the condition is treated by the use of some bland antiseptic ointment (e.g. sulphacetamide 6 per cent) combined with a firm pad and bandage. The latter prevents movements of the lids over the abraded area and allows the epithelium to cover the defect. Most abrasions heal within forty-eight hours, but infection and failure to heal will lead to corneal ulceration.

Recurrent Corneal Erosion

A variable period after some small corneal abrasion, and when the condition is apparently firmly healed, the patient may wake one morning with a painful watering eye and a staining area at the site of the original lesion. After treatment with ointment and a firm pad, the condition once more heals, only for the history to be

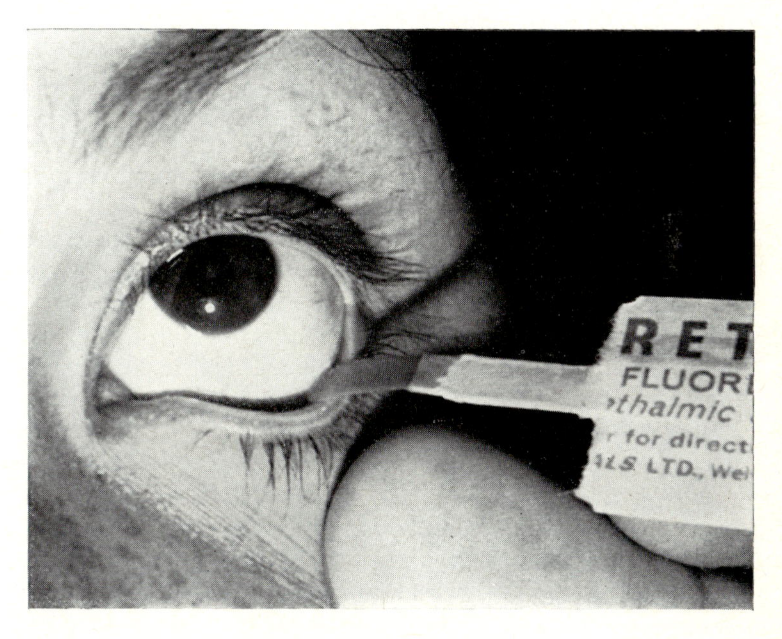

Fig. 21 Staining the cornea.

repeated after a further interval. It seems that the epithelium, after the original lesion, fails to become firmly adherent to its bed and is liable to be torn off by the first opening of the lids in the morning. It is probable that some low-grade virus (perhaps allied to herpes) gains access to the epithelium and prevents complete healing.

The first line in treatment in hospital is the removal of all the loose epithelium with absolute alcohol (after anaesthetising the eye). After the epithelium has once more covered the defect the patient is advised to use some lubricant to the eye each night for a period of several weeks, in order to allow complete adherence between the epithelium and Bowman's membrane. Either liquid paraffin or sulphacetamide ointment are suitable for the purpose.

Foreign Bodies

In an industrial area corneal foreign bodies are very common as they mostly arise from machining processes, especially grinding and polishing, but a dry windy day will lead to the condition in motorists, cyclists and pedestrians.

In most cases there is a definite history of something having entered the eye and of an acute onset of pain, especially on blinking, though it is remarkable how often a workman will fail to notice the actual entry of the particle into the eye, only becoming conscious of it after some hours. The foreign body can usually be seen lying on the cornea or embedded in the surface, but some particles, like seed husks or wings of insects, may be so small and translucent that their presence cannot be excluded until the eye has been examined in a good light and with some degree of magnification. It may be necessary to stain the cornea with fluorescein for this. Adequate illumination, again, is essential if the removal of these fragments is to be undertaken with ease and the minimum of damage. The eye is anaesthetised with a drop or two of amethocaine 1 per cent, and the foreign body is picked off the surface with a sharp instrument. If a needle mounted in a handle is not available, a large hypodermic needle will serve very well. Providing the position and depth of the particle have been accurately assessed there need be no hesitation felt if a certain amount of 'excavation' is required to free a deeply embedded particle. There is virtually no risk of penetrating the cornea with the needle. The traditional, blunt-ended 'eye-spud' has no place in the removal of corneal foreign bodies. Its use will only lead to further damage to the epithelium.

It will be found that many foreign bodies can be removed by simpler means than the above; some will, in fact, simply wipe off the cornea. Pieces of metal, especially those which have been present for a few days, are often surrounded by a ring of rusty staining. An attempt should be made to remove all of this as its presence tends to irritate the eye. If the first attempt to do this is not successful, it will be found that the rust becomes loose after a day or two, and that it can then be removed.

It must not be forgotten that, if a patient complains that there is a foreign body in the eye and it is not seen on the cornea, search must be made for some other irritant. There may be a misplaced eyelash, or a foreign body beneath the upper lid.

Following the removal of a foreign body from the cornea, the eye should be covered for a day or two and protected from infection by the use of an antiseptic ointment. It must be remembered that the commonest cause of corneal ulceration is infection of a corneal abrasion or other breach in the epithelium. If the particle has been present for more than a few hours, the eye may be appreciably red and irritable, and a drop of atropine may be needed. Dilatation of the pupil in this way is, however, unnecessary after the removal of most foreign bodies and should be avoided if possible. As the effects of atropine take a week to ten days to wear off, the disability resulting from the use of this mydriatic is quite considerable.

Welders Flash

This condition, also known as 'arc flash' is similar in pathology to snow blindness, and is due to exposure to ultraviolet light. The patient is usually a welder or a welder's mate who has been unguardedly exposed to a welding arc. After an interval of six or eight hours, intense bilateral lacrimation, blepharospasm and photophobia develop and the patient is abjectly miserable.

Examination of the eyes is impossible until the eyes have been anaesthetised with 1 per cent amethocaine. Once the blepharospasm has been relieved by this and the eyes stained with fluorescein it is seen that the cornea is stippled with innumerable punctate erosions. These lesions heal within a few hours and the relief of the blepharospasm by the local anaesthetic, together with reassurance that all will be well, is usually the only treatment required. If discomfort returns when the effects of the anaesthetic wears off, cold compresses are soothing. The entire episode is over in a few hours.

Perforations of the Cornea

These are always serious injuries demanding immediate attention, as they are commonly associated with damage to the deeper parts of the eye and may lead to intraocular infection.

Penetrations may be in the form of frank ruptures, due to large foreign bodies or to severe contusion injuries, or they may be so small as to be almost invisible, particularly when situated at, or close to, the limbus. In the latter case the injury is due to a small flying fragment, often of metal, and this will often be retained within the eye, though needles, wire and other sharp instruments cause similar injuries.

In the diagnosis of gross injuries no difficulty is likely to be experienced, provided the eye is examined in a good light, and steps are taken to relieve the invariable blepharospasm by local anaesthesia. Small penetrations require close examination for their detection and the possibility of such an injury must be constantly borne in mind. The history is of prime importance as the majority of these injuries occur as the result of processes in which flying fragments of metal are liable to be thrown off. Various industrial processes are examples in point, and the worker will often have failed to wear protective goggles. Many penetrating injuries of the eye, however, occur as a result of the use of everyday tools, notably the hammer and chisel. If a chisel is not maintained in good condition by regular attention on the grindstone, its edges become cracked and small flakes are knocked off by the hammer. It is these flakes, or pieces of the hammer itself, which enter the eye at high speed, often through a minute penetration.

The possibility of penetration should, therefore, be constantly in mind if the injury has occurred during processes such as have been described.

The vision should be estimated and the outer parts of the eye inspected carefully. A small opacity in the cornea may be noted or there may be disturbance of the iris (bleeding into the anterior chamber, tears of the iris substance or interference with pupil function). Examination of the deeper parts of the eye, after dilating the pupil with homatropine 1 per cent if necessary, may reveal opacities in the lens or vitreous, or haemorrhage within the eye. Any of these features may be evidence of a serious injury necessitating specialist attention. If there is doubt as to the presence or absence of an intraocular foreign body the decision will be made after X-ray examination of the orbit. The importance of early attention to these injuries lies in the fact that an eye which retains a ferrous metallic fragment within it is invariably lost eventually, on account of the chemical changes which take place between the iron and the intraocular fluids (*Siderosis bulbi*).

With modern treatment, many eyes which have sustained a penetrating injury can be saved, often with useful vision, particularly if the lens is undamaged. Lacerated wounds can be repaired with fine sutures and magnetic foreign bodies removed with a magnet. A delayed complication, which sometimes spoils a successful result, is detachment of the retina due to contraction of the scar of the injury or to organisation of intraocular haemorrhage.

CORNEAL INFLAMMATION (KERATITIS)

Ulcerative Keratitis

In corneal ulceration, although the lesions vary considerably in their nature and specialised nomenclature, they have one feature in common—a breach in the corneal epithelium. The most important single weapon in their diagnosis, therefore, in addition to a good light, is fluorescein which, as has been said, is taken up by the damaged corneal epithelium and shows up as a bright green stain. While no attempt to describe all the types of ulcer is proposed, some typical varieties will be discussed. The general lines of treatment are the same for them all.

Simple Corneal Ulcer

This most commonly arises from infection entering a corneal wound or abrasion and there is a history of injury; or entropion, with rubbing eyelashes, has abraded the epithelium. The eye is irritable, photophobic and watery, while the patient complains of a sensation as of something in the eye. This is due to the movement of the lids over the damaged epithelium.

On examination the eye is red and this redness is most marked around the margins of the cornea, particularly in the quadrant where the ulcer lies. Vision is usually reduced, often markedly so, if the ulcer occupies the central area. Even before the stain is used a greyish area of oedema and infiltration is to be seen in the cornea. Fluorescein will demarcate the edges of the ulcerated area. A thorough search should be made for a foreign body, not only on the cornea, but also in the conjunctival sac and under the upper lid. If the ulcer is small it will often heal as a result of treatment with some chloramphenicol ointment and the application of a firm pad. A more severe inflammation will require the addition of atropine (1 per cent drops or ointment) twice daily in order to dilate the pupil and to put the intraocular muscles at rest, for any severe corneal lesion is associated with some degree of iritis. Hot bathing is also a comfort.

If the ulcer is extensive and necrotic, with much debris in the crater, it should be sterilised by direct application of antiseptics in the following way. The eye being thoroughly anaesthetised with amethocaine, the cornea is dried with blotting paper and the ulcer is touched with a sharpened matchstick which has been dipped in pure phenol or absolute alcohol, particular attention being paid to the edges of the lesion. It is helpful for an assistant to retract the eyelids. The cornea must be dry during application of the caustic. A drop of atropine and some ointment are instilled and the eye is bandaged. A resistant ulcer will usually respond to this treatment, which is most likely to be carried out in a hospital department, where special facilities are available. If hospital facilities are not available, cauterisation of a corneal ulcer in this way can be regarded as a surgery procedure.

Hypopyon Keratitis

If the corneal ulcer is a severe one and the invading organism virulent or the patient debilitated, the reactionary iritis is intense and there is an outpouring of cells into the anterior chamber of the eye. These cells settle to the lowest part and form a collection of pus (*the hypopyon*) (Fig. 22). This pus is sterile and remains so unless the ulcer becomes so deep as to perforate the cornea, in which case organisms enter the eye from outside. Hypopyon formation is most common after injury by stone and coal and is particularly prone to occur in miners. In a proportion of cases the patient is found to have an obstructed and infected lacrimal sac on the affected side. Removal of this source of infection is essential to the success of treatment.

Since the introduction of antibiotics into the treatment of corneal injuries at the earliest possible moment (in First Aid Rooms) the incidence of hypopyon keratitis has fallen dramatically, but it is still an important cause of visual defect and loss of working time. These severe ulcers cannot be treated domestically and the majority require in-patient treatment in hospital. In addition to the treatment used for corneal ulcers in general, systemic use of penicillin is often called for, and application of antibiotics to the eye. These are used as drops and also by subconjunctival injection. After anaesthetising the eye thoroughly, framycetin 500 mg dissolved in sterile water can be injected beneath the conjunctiva and this is often given at the same time as 5 minims of Mydricaine, a combination of atropine, adrenaline and cocaine, acting together as a powerful mydriatic.

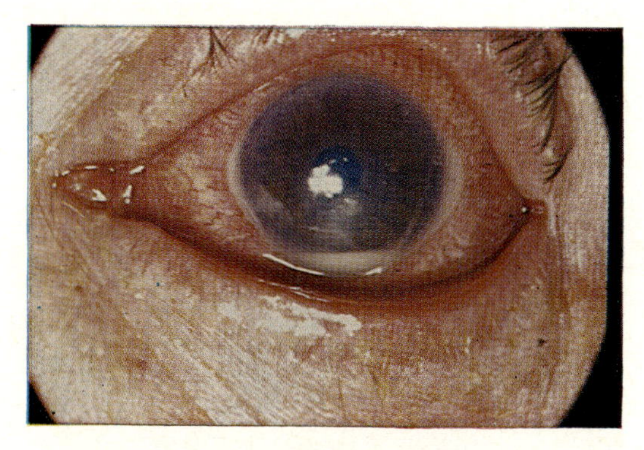

Fig. 22 Hypopyon ulcer.

Herpes Simplex (*Dendritic Ulcer*)

This is a variety of ulcer which, if secondary infection does not occur, is confined to the epithelium of the cornea and leaves only a faint and often non-permanent scar behind it. The condition often appears during or after some mild febrile upset or a 'cold in the head', and is due to the invasion of the epithelium by the herpes virus. An attack may also be triggered off by exposure to bright sunlight, or by emotion. A particular feature of dendritic corneal ulceration, at least in the early stages, is the relative lack of irritation and photophobia. The eye often remains quite white. This effect is thought to be due to the depression of corneal sensitivity resulting from herpetic infection.

When stained with fluorescein the ulcer has a typically branched form with small nodules or vesicles at the ends of the branches (Fig. 23).

Removal of all unhealthy epithelium and sterilisation of the affected area by painting with absolute alcohol is the first step in the treatment, and will be carried out by the specialist, though the general practitioner should not hesitate to carry out the treatment himself if the need arises. Afterwards the eye is treated on the usual lines, with a firm pad, soothing ointment (chloromycetin is especially useful), and atropine if the eye is very irritable. The condition shows a tendency to recurrence, and may lead ultimately to severe visual defect from corneal scarring. Corneal grafting is being increasingly used in an effort to prevent recurrences and to improve vision.

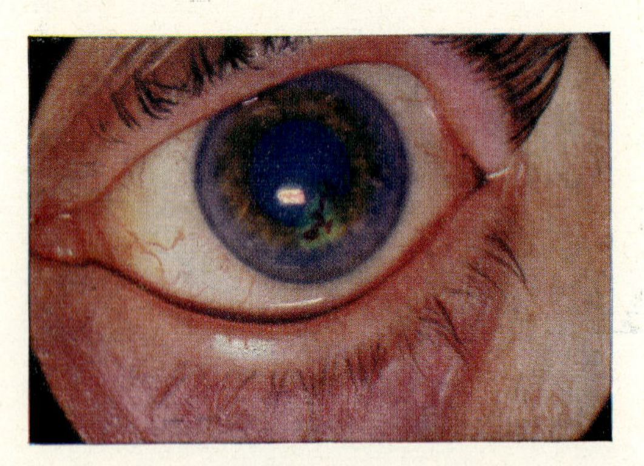

Fig. 23 Dendritic ulcer of cornea.

The occurrence of dendritic ulceration is an absolute contra-indication to the use of cortisone and similar preparations by practitioners without expert guidance. Even when combined with antibiotics, corticosteroids may aggravate lesions of this nature.

The recent introduction of antiviral agents may alter the treatment of herpetis keratitis, particularly in these cases where the condition is confined to the epithelium with no involvement of the corneal substance. 5-iodo-2-Deoxyuridine (IDU, Kerecid) is the most frequently used of these drugs. In an early case the frequent use (hourly) of IDU drops, combined with padding of the eye, for a period of at least five days, may lead to healing of an epithelial lesion and avoid the necessity of chemical cauterisation. This may be of particular value in the case of a child when alcoholisation of an ulcer will require a general anaesthetic.

Superficial Punctate Keratitis

A condition which is probably related to herpetic keratitis and which leads to long-continued mild symptoms of irritation and which has a tendency to spontaneous resolution. The patient has a little blepharospasm and lacrimation and examination after staining with fluorescein shows a number of punctate staining points in the corneal epithelium. Treatment with chloromycetin ointment two or three times a day is a comfort.

Marginal (Catarrhal) Ulcer

In middle-aged and elderly individuals, recurrent episodes of inflammation involving the margin of the cornea occur. Often with little associated conjunctivitis, but with vascular congestion at the junction of cornea and sclera, small creamy infiltrates (single or multiple) are to be seen at the very margin of the cornea, and *never* progressing centrally.

It is thought that these small lesions, which may or may not stain with fluorescein, represent an allergic response. If they do not rapidly heal as a result of treatment with antibiotic drops locally, a short course of local betamethasone will produce a cure. At least until the next attack!

NON-ULCERATIVE (INTERSTITIAL) KERATITIS

Although the term interstitial keratitis traditionally refers to a specific condition, it is also technically applicable to a number of inflammatory corneal lesions in which the epithelium is intact (as shown by the fluorescein test) and in which the inflammatory process occurs in the deeper layers of the cornea.

Disciform Keratitis

This is another form of 'virus keratitis'. It has an acute onset, with a red photophobic eye and diffuse corneal haze. After a number of days the haziness, which is due to oedema, becomes localised in the centre of the cornea and it is seen that this opaque disc is appreciably thickened. After a period of weeks the inflammatory oedema settles down and the patient is left with a permanent corneal opacity of varying density. Considerable visual disability results from this.

There is a close relationship between disciform keratitis and dendritic ulceration. Particularly is this the case where a dendritic ulcer has been mistakenly treated with steroids when the virus may, as it were, be driven into the deeper layers of the cornea with the production of an interstitial lesion.

Until recent years treatment was unsatisfactory and consisted of the routine use of atropine, combined with hot bathing during the period of acute inflammation. The advent of steroids has altered the picture to some extent and disciform keratitis is one of the conditions in which the local use of steroids may have a beneficial effect. The hormone acts by reducing the oedema and thereby limits the final corneal opacity. If this opacity is severe, corneal grafting holds out hope of visual improvement.

The fact that local steroid therapy may be indicated in disciform keratitis while absolutely contra-indicated in dendritic ulceration, increases enormously the difficulties of planning the correct treatment of these two conditions. It can only be emphasised that much greater risks are attached to the improper use of steroids locally than to their being withheld. If there is any doubt as to the correct course to adopt, steroids must *not* be used, and reliance should be placed on atropine and chloromycetin.

Specific Interstitial Keratitis

Occurring in patients with congenital syphilis and now relatively rare, interstitial keratitis is a bilateral disease of acute onset and most often seen in girls between the ages of five and seventeen. There may be a trivial injury as a precipitating factor.

The scarring resulting from specific interstitial keratitis is one of the conditions in which corneal grafting is indicated. The results are very good and dramatic visual improvement may result.

Keratitis Occurring During the Course of Other Diseases

Corneal affections are not uncommon during the course of systemic disorders. Among the more important conditions are the following:

1. ACNE ROSACEA. Vascularising keratitis is the most disabling complication of rosacea. It is most often seen in middle-aged women and is bilateral. Exacerbation of the eye condition does not seem to be necessarily dependent on a relapse of the skin condition, but may occur independently of this.

The corneal affection is superficial and at first takes the form of marginal opacity which spreads on to the cornea and is vascularised from the limbus. Recurrent attacks lead to further scar formation and the final visual defect may be considerable, so much so that corneal grafting may be called for, in an attempt to remove the scarred area of the cornea.

Local treatment is purely symptomatic as we know of no means to prevent recurrences. Mydriatics help to make the patient more comfortable. Acne rosacea keratitis is another of the corneal conditions in which steroids are being used locally to keep an inflammatory attack within bounds while waiting for natural remission to take place. The hormone is of great benefit, but it has no effect on the fundamental metabolic upset which is the basis of the condition, the nature of which is unknown.

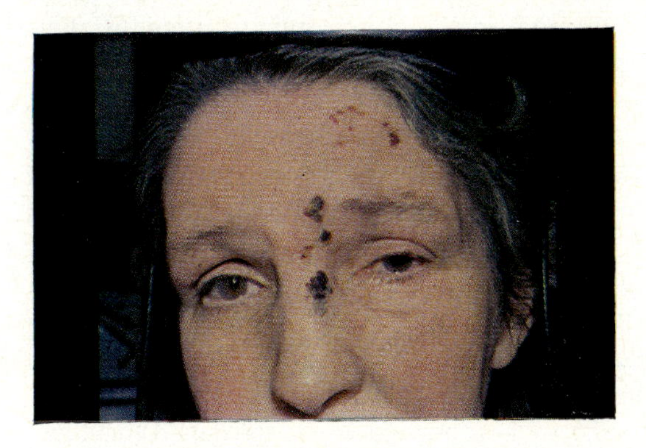

Fig. 24 Herpes zoster ophthalmicus.

2. HERPES ZOSTER OPHTHALMICUS (Fig. 24). A proportion of cases of shingles affecting the first division of the fifth nerve develop ocular involvement. The degree of this involvement may range from a trivial conjunctivitis to a devastating kerato-conjunctivitis, and perhaps secondary glaucoma. On the subsidence of the acute process, the patient may be left with an insensitive cornea and be exposed to the risks of neuroparalytic keratitis (q.v.).

It is usual for the eye signs to appear at the same time as the skin vesicles, but it must be remembered that the eye may be affected severely while the skin lesions are minimal, or the ocular condition may overshadow the basic pathology. It is classically taught that vesicles at the tip of the nose, due to involvement of the external nasal nerve, imply a greater degree of risk to the eye than is present in those cases in which this sign is absent.

It is the extreme variety of the clinical picture that makes description of the 'typical' case difficult. If the ocular involvement is of some severity, the eye will be photophobic and irritable, with redness mainly at the corneal margin. Patches of opacity will be seen in the corneal substance and a degree of iritis will be manifest by a sluggish reaction of the pupil and perhaps some distortion of its margin. Corneal sensitivity (as tested with a wisp of cotton wool) may be diminished. Owing to the lack of sensation, ocular complications, especially secondary glaucoma, may develop relatively painlessly. The best clinical guide to the need or otherwise for specialist advice is an assessment of the visual acuity. If this is depressed, even though inflammation of the eye is not marked, specialist help is called for.

Such a condition is best treated with atropine (1 per cent b.d.) together with chloromycetin ointment, while a daily watch is kept on the integrity of the corneal epithelium. If the fluorescein test shows that the epithelium is no longer intact, there is a danger of frank ulceration developing and specialist advice is required. This danger is increased in the presence of diminished corneal sensitivity, as a functioning nerve supply plays a great part in the maintenance of corneal nutrition. If a firm pad and bandage does not prevent spread of the ulceration, tarsorrhaphy will probably be necessary, the lids being allowed to remain together until some months after the disease is apparently healed.

3. NEURO-PARALYTIC KERATITIS. This follows interference with the fifth cranial nerve and, through it, with the nutrition of the cornea which depends, as has been said, partly on the integrity of its nerve supply. Neuro-paralytic keratitis is, therefore, a possible complication of head injuries and of intracranial operations, particularly trigeminal root section for trigeminal neuralgia.

Sensitivity of the cornea is defective, there may be spots of interstitial opacity, and there is a risk of severe corneal ulceration developing. This is due to the fact that abrasions and other trivial injuries pass unnoticed and the usually painful results of inflammation are absent. Despite the fact that the cornea remains insensitive, the risk of the development of serious lesions diminishes as time passes and the cornea adapts itself to the new conditions. In the early stages, therefore, the eye must be protected from accidental injury by the fitting of a specially designed goggle carrying a side-piece or transparent 'blinker' shaped to the temple. After six months this can be discarded with safety.

4. EXPOSURE KERATITIS. The integrity of the corneal epithelium depends in part on its surface being kept moist by the constant movements of the lids. If this action is not present, or is prevented from taking place, drying of the cornea leads to erosion and consequent ulceration. Among the many conditions which may lead to this are destructive lesions of the lids after burns or trauma, gross exophthalmos in the course of thyrotoxicosis or orbital tumour, facial palsy, and severe debilitating diseases, such as pneumonia or advanced heart failure, where the eyes remain partly open for long periods and blinking does not occur. It is the lower third of the cornea which suffers from the exposure as the upper part is protected by the lid and by the natural tendency for the eye to be rolled upwards on attempted closure. The first sign of exposure keratitis is a lack of the normal lustre of the epithelium and this is followed by frank loss of tissue and infection, with ulcer formation and staining with fluorescein. If neglected, corneal perforation may occur and the eye be lost. This is a preventable condition and the possibility of its occurrence must be borne in mind. If the cause of the exposure is likely to be short-lived, twice daily lubrication of the eye with an abundance of bland antiseptic ointment is all that is required. In the case of lid injuries, or of prolonged exophthalmos, partial or total tarsorrhaphy is the only safe course. In the treatment of the established condition, the general principles of the treatment of corneal ulceration apply—atropine, antiseptic ointment, and a firm pad and bandage to keep the eyelids closed.

DEGENERATIVE CONDITIONS

Band-shaped Opacity

In some degenerate eyes, and in eyes which have sustained injury or ulceration in the past, a transverse band of opacity develops in the lower part of the cornea and this may be associated with the deposition of white calcium salts. Irritation is caused by the breaking free of flakes of this deposit and the production of defects in the epithelium. Treatment is unsatisfactory, but removal of the chalky plaques is sometimes comforting and some of the opacities respond to specialised chemical treatment.

MATERIAL FOR CORNEAL GRAFTING

The only material that has been proved satisfactory for the surgical replacement of the opaque cornea is fresh corneal tissue.

There are two practical sources of supply: (1) from eyes removed on account of some disease in the deeper parts of the eye, usually intraocular tumour; and (2) from eyes removed shortly after death. It is with the second group that the practitioner is likely to be concerned.

In Great Britain eyes will be used for this purpose if the person has expressed a wish for this to be done. This wish need not be in writing but should be made clear to the next-of-kin or those likely to be in the house at the time of death, and, preferably, to the general practitioner. As soon after death as practicable, the practitioner will communicate with the nearest Eye Department, where arrangements will be made to collect the eyes. This should probably be done within about six hours after death. Under suitable conditions of storage, the cornea can be used for grafting up to a week afterwards.

Another possible source of donor material is from people dying in hospital, when the authorities are entitled to remove the eyes, providing the relatives have no objection. The difficulty here is with regard to the delay which may occur between the time of death and the opportunity to discuss the matter with the relatives.

CHAPTER SIX

Intraocular Inflammation

ANATOMY

The vascular coat of the eye, which serves to nourish the retina and the other intraocular structures, comprises, from before backwards, the iris, the ciliary body and the choroid. Its blood supply is derived from branches of the ophthalmic artery which pierce the sclera at the posterior pole of the eyeball close to the optic nerve head and also, in front, in relation to the insertions of the external ocular muscles. In addition to its nutritive functions, the ciliary body is specially modified to take part in the mechanism of accommodation by means of the muscles which it contains, and also to produce, through the ciliary processes, the aqueous humour, the circulation of which is so important, not only in the maintenance of the nutrition of the avascular structures of the eye, particularly the lens, but also in the maintenance of the intraocular pressure.

UVEITIS

It is convenient to consider inflammatory diseases of the uveal tract as one group, for this serves to remind us that the three parts of this vascular tract are continuous one with the other.

Aetiology. Many local and systemic factors have been implicated as possible causes of intraocular inflammation and any simple classification is very difficult. There are, however, three fairly well defined varieties. These are: (1) exogenous, (2) secondary, (3) endogenous, or of unknown aetiology.

Exogenous Uveitis

These cases include those in which infection has been introduced into the eye from without and it follows, most typically, infection of an operation wound, or infection carried in by penetration of

For Anatomy of the Eye see Figure 1.

the coats of the eye by a foreign body. This type of disease may also follow the penetration of a corneal ulcer or the breaking down of the cornea following some degenerative condition.

Secondary Uveitis

Here inflammation of the uveal tract results from such conditions as:

1. Keratitis.
2. Retinal detachment, particularly of long-standing.
3. Intraocular new growth.
4. Haemorrhage within the eye.
5. Trauma.
6. Sympathetic ophthalmia. (This condition follows a penetrating injury or operation to one eye and leads to iridocyclitis in the second eye. It is discussed in more detail during the consideration of injuries to the eye.)

Endogenous Uveitis

In this group must come the majority of cases of intraocular inflammation, the aetiology of which in most cases is problematical. There are one or two conditions, such as ankylosing spondylitis, sarcoidosis and diabetes, which are known particularly to be associated with inflammation of the eye, and uveitis sometimes occurs in association with generalised febrile illnesses. It is also seen in malaria, sometimes in virus disease, and as part of one or two somewhat obscure syndromes manifest as dermatological or systemic upsets, associated with uveitis. Toxoplasmosis and brucellosis have been implicated in some cases.

It must, however, be admitted that despite an intensive investigation into the patient's general condition with X-rays, sensitivity tests, blood tests, and detailed clinical examination, in the majority of cases it is not possible to identify any definite aetiological factor. The impression is, however, that these unexplained cases of uveitis are of the nature of a sensitivity reaction, perhaps to some allergen or systemic toxic process, and that clinical research will gradually enable us to classify more and more cases of uveitis into groups having a discoverable causative factor. At the moment these conditions are treated somewhat empirically and symptomatically, and it is a fortunate feature of the disease process that most cases seem to be self-limiting and not to suffer from a severe recurrence. There are still the less common cases in which, despite all efforts to find a causative factor, recurrent attacks of inflammation within the eye lead to increasing interference with function, increasing opacity of the ocular media, the development of cataract, and often irrecoverable visual defect.

Clinical Features. It is not possible to describe the many variations in the clinical presentation of uveitis, but it should be remembered that the condition may range from an acute fulminating exudative process to a low grade chronic condition in which gradual visual failure may be the patient's only complaint. It is proposed, therefore, to discuss the acute inflammatory process as it affects the uveal tract, subdividing it into only: (1) anterior uveitis (i.e. *iritis* or *iridocyclitis*), (2) posterior uveitis (*choroiditis*).

Acute Iridocyclitis (Anterior Uveitis)

The patient suffering from anterior uveitis is usually young. He may apparently be in a good state of general health, when, without any premonitory symptoms, he develops pain in an eye. The eye is tender to the touch, bright light is irritating and there is excessive lacrimation. Any attempt to do close work increases the discomfort, and visual acuity is diminished. This diminution of visual acuity varies from case to case, but it is not usually extreme, a feature which distinguishes iritis from an attack of acute congestive glaucoma.

On examination the eye is red, the redness being located mainly at the margin of the cornea (*ciliary injection*). This is due to congestion of the anterior ciliary blood vessels where they are lying close behind the limbus. The cornea is bright and does not stain with fluorescein, but the view of the iris is partially obscured by oedema and sometimes by frank exudation into the anterior chamber of the eye. This oedema of the iris is manifest by lack of definition of the normal pattern of the iris.

The pupil is somewhat small. It reacts poorly to light, and, if the condition is of more than a few hours duration, it will usually be seen that adhesion of the iris to the anterior surface of the lens has distorted the pupil, which, instead of being round, is now irregular and on attempted dilatation with drugs dilates poorly at the site of its attachment of the lens (Fig. 25). Even though regular dilatation of the pupil may be obtained with mydriatics, it will often be seen that there has been left behind a deposit of pigment on the anterior surface of the lens capsule, this pigment being derived from the inflamed iris. If it is the ciliary body which is playing a greater part in the inflammatory process than is the iris, then it may be possible to see clumps of lymphocytes adherent to the posterior corneal surface. These cells have escaped into the aqueous humour from the inflamed ciliary body and have been carried by the circulating aqueous into the anterior chamber, where they adhere to the posterior surface of the cornea,

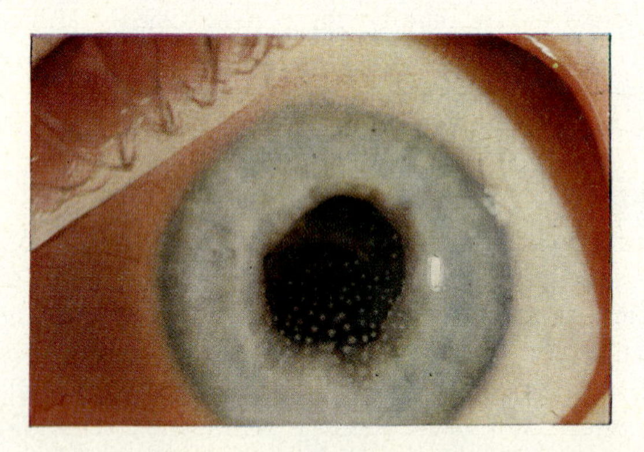

Fig. 25 Iris adhesions—synechiae.

usually in its lower part. Though very difficult to see in their early stages, these *keratic precipitates* (K.P.) (Fig. 26) may become of quite an appreciable size and they will form definite evidence of involvement of the ciliary body. When the active phase of the disease is at an end, the K.P. gradually become crenated and pigmented and usually remain as a permanent record of antecedent iridocyclitis.

There is one final feature of anterior uveitis which should be mentioned; that is the tendency of inflammatory cells to find their way, not only into the anterior chamber with the production, as has been shown, of keratic precipitates, but also into the posterior part of the eye. The passage of cellular and protein rich material into the vitreous leads to the increasing opacity of the vitreous body. If of minor degree, this opacity may show itself simply as fine threads or clumps of material obstructing the clear ophthalmoscopic view of the retina and seen by the patient subjectively as 'spider's webs' and 'floating particles'. If of greater degree, vitreous opacities of this nature may become a serious impediment to vision.

As has been said, the picture of an acute inflammatory process may be replaced by a low grade chronic recurrent inflammation in which there are no features drawing the patient's attention to serious disease. It may only be after visual defect becomes severe or after an abnormality of pupil function is noticed that inflammatory corneal deposits are seen to be present and vitreous opacities developing.

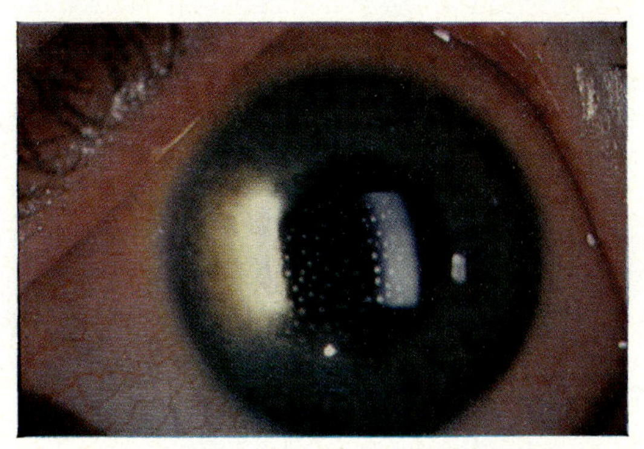

Fig. 26 Keratic precipitates—K.P.

Treatment. This is based on three main principles: (1) relief of pain, (2) prevention of adhesions between the inflamed surfaces and their separation if already formed, (3) search for the cause.

Pain is relieved most effectively by the use of mydriatic drugs which have the added value of breaking down or preventing iris adhesions. The paralysis of iris and ciliary body which results from the use of mydriatics also relieves pain by the abolition of the irritating dilatation and contraction of the pupil and the pull of the ciliary body during the movement of accommodation. The most useful mydriatic is atropine and it is used in the strength of 1 per cent drops or ointment thrice daily in the early stages. When the inflammation is under control the frequency of the application can be reduced.

Atropine is combined with the application of heat to the eye. This may be by simple hot bathing or by more elaborate measures such as medical diathermy. In addition analgesics may be required to relieve pain and sedatives to ensure sleep.

Steroids may be used as local applications in acute inflammatory conditions in the anterior part of the eye. They act by reducing the exudative inflammatory process and are in no way curative of the underlying condition, but, as the usual variety of iridocyclitis is self-limiting, cortisone and its derivatives play a valuable part in the reduction of exudate and, therefore, in the reduction of the residual visual defect.

One of the complications of iridocyclitis is glaucoma, and an inflammatory cause has to be considered when we are investigating a case of glaucoma which does not fit into one of the so-called 'primary' groups. If glaucoma supervenes in a case of iridocyclitis, the patient will complain of a sudden increase in the visual defect with increased pain in the eye. These complaints indicate that some secondary condition has supervened and point to the need for specialist advice.

In chronic iridocyclitis, interference with the nutrition of the lens leads to the development of cataract. When the inflammatory process is at an end, surgical removal of the opaque lens may lead to visual improvement.

Posterior Uveitis (Choroiditis)

This condition, even in its acute form, is painless, though the patient will sometimes complain of a little aching and discomfort in the eye. Visual defect is the only complaint, and the degree of this depends upon the site of the lesion. A patch of choroiditis close to the macula will produce an immediate complaint of defective vision, but a patch further towards the periphery may be unnoticed until the vitreous opacity resulting therefrom is of sufficient density to interfere with vision. Some patients in fact are found on routine ophthalmoscopic examination to have the scars of healed choroiditis but to be unaware of the occurrence of any acute phase.

The clinical findings, apart from diminished visual acuity, depend mainly upon examination with the ophthalmoscope, and dilatation of the pupil may be required before this examination can be adequately carried out. Careful examination of the cornea will show clumps of inflammatory cells (keratic precipitates) on the posterior corneal surface. They are evidence of the inflammatory reaction spreading throughout the intraocular fluids. In a well-developed case detailed examination of the retina is obscured by a haze in the ocular media. This may be quite well localised in one region of the fundus, but it may also be more diffuse. It is due to the exudative opacity of the vitreous which has already been described. On examination of the posterior part of the eye, it will be seen that there is a greyish hazy patch with somewhat ill-defined margins in one quadrant of the fundus, perhaps close to the optic nerve head. Retinal vessels in the surrounding region may be poorly seen as they are concealed by the retinal oedema consequent on the patch of choroidal inflammation. With resolution of the inflammatory process, the patch of choroiditis becomes more well-defined, it develops sharp irregular margins, and proliferation of the retinal pigment cells leads to a patchy dark discoloration of the scar. The scar itself is white, due to the fact that atrophy of the choroid in the area has occurred, allowing the bare sclera to be seen with the ophthalmoscope.

The visual defect due to the presence of the scar is, of course, permanent, and its importance from the patient's point of view depends once more on its relation to the macula. With resolution of the inflammation the vitreous tends to clear and the visual defect resulting from vitreous opacity becomes less severe as time passes. The condition, in common with other varieties of uveitis, is usually of obscure aetiology and prone to recurrence. Current opinion ascribes an increasing number of cases of choroiditis to infection with toxoplasma, and another parasite more recently implicated, especially in children, is the Toxocara. This is apparently ingested by a child who is in contact with a dog carrying the parasite. It is suggested that puppies should be 'wormed 'regularly.

Treatment. Here, local treatment is less likely to be of value than in anterior uveitis. Drugs applied to the anterior part of the eye do not reach the vitreous chamber or the choroid, and atropine is indicated in choroiditis only if there is evidence of coincident involvement of the ciliary body. This involvement will be shown by the presence of keratic precipitates on the posterior surface of the cornea. The handling of a case of posterior uveitis is an extremely difficult problem, but it seems that the patient should be confined to bed or at least to a rigid regime of rest during the early stages. The use of corticosteroids systemically is now being considered in this condition in an effort to reduce the exudation into the vitreous, to minimise the ultimate size of the resultant scar, and to reduce the effects of the retinal oedema on the visual functions. If the patch of inflammation lies close to the macula and threatens central vision, it seems that the use of one of these drugs is justified.

INVESTIGATION OF THE CAUSE OF UVEITIS. The cause of uveitis in a given case is almost always obscure, but it is usual for the patient to be investigated from the general medical point of view lest there should be found some definite source of sepsis, treatment of which might be indicated in an effort to prevent recurrence of the uveitis. With this end in view, detailed physical examination is undertaken, X-rays of chest, sinuses and teeth are examined, together with X-ray examination of the lumbar spine to exclude ankylosing spondylitis. Mantoux reaction is tested and blood is taken for Wassermann reaction. The usual result of these investigations is on the whole disappointing, but there is the occasional case in which some specific disease is discovered, treatment of which would seem, on general grounds, to be desirable.

CORTISONE AND ALLIED DRUGS

THEIR USE AND ABUSE

The availability of cortisone, hydrocortisone, prednisolone and other steroid drugs for local use has radically altered the treatment of some eye diseases, particularly of the intraocular inflammations, iritis and iridocyclitis, but experience shows that the use of these drugs is not without danger.

The local use of steroid drugs reduces the inflammatory response, and there is a great temptation to use them in the case of any red eye. They can be almost guaranteed to produce rapid decrease in congestion and discomfort, but this apparent improvement may be illusory and conceal the damage taking place.

The principal effect of steroids in the eye is to reduce the inflammatory reaction, to limit exudation, and to decrease congestion. In some cases this effect is purely beneficial and these drugs have their greatest use in the control of iritis and iridocyclitis, where exudation and a damaging vascular reaction can lead to permanent interference with vision. In other cases, however, reduction of the vascular response hinders the body's repair mechanism, and this is especially applicable in the case of the cornea. There is no doubt that cortisone and its allies have an ill-effect on the healing of corneal tissue in many circumstances.

Even though cortisone and hydrocortisone are now available in combination with various antibiotics, it is considered that their use is absolutely contra-indicated in the majority of corneal ulcers and in inflammatory diseases of the conjunctiva. There are one or two exceptions to this rule, particularly in the case of rosacea keratitis and spring catarrh, but the decision as to whether or not these drugs should be used has often to be agreed on rather technical grounds and as a result of specialised methods of examination.

It is in the case of herpes simplex of the cornea (dendritic ulcer) that the greatest danger of the steroids lies. Despite the combination of the drugs with antibiotics, a dendritic ulcer treated with steroids invariably adopts a virulently destructive character, spreading through the corneal epithelium very rapidly and often producing permanent scarring.

Statements have been made to the effect that corticosteroids are safe if combined with one of a variety of antibiotics. Many experienced ophthalmologists now hold that such statements do a great disservice to the profession and, often, to the patient.

It must not be considered that this risk is being overemphasised, for reports from eye hospitals indicate an increasing awareness of the danger on the part of the eye specialists and quote case histories in which the ill-advised use of steroids has led to increasing damage to the eye.

The general practitioner is faced with a very difficult problem in many cases of 'the red eye'. He will often not have the knowledge and experience to decide the exact nature of the condition with which he is faced, and yet there is a great temptation to use a drug which will reduce the inflammation of the eye and make the patient more comfortable.

Ideally, the steroids should not be used in the eye without there being a definite diagnosis, and if this diagnosis cannot be made with certainty then specialist help should be sought, or less dangerous remedies employed. Certainly, if there is any doubt as to the integrity of the corneal epithelium, steroids should not be prescribed.

It should be noted that the long-term use of local steroid drops occasionally gives rise to glaucoma. This is another reason why these drugs should not be used over long periods without supervision by an ophthalmologist.

The use of prednisolone systemically over a long period occasionally leads to the development of bilateral cataract.

CHAPTER SEVEN

Cataract

CLASSIFICATION OF CATARACT

1. Congenital.
2. Senile.
3. Complicated (i.e. arising in the course of some other disease of the eye).
4. Diabetic.
5. Traumatic.
6. Other types (this category includes endocrine cataracts, cataracts due to toxic hazards or radiation, and cataract due to poisoning).

It must be remembered that any opacity in the lens of the eye is technically a cataract, and the term cataract should, therefore, be used carefully in the presence of patients, as it still conveys visions of a painful operation, a long period of partial blindness and uncertain cure. Many cataracts, however, are non-progressive or slowly developing, and it is only a small proportion that need treatment by the surgeon.

CONGENITAL CATARACT

The lens is an ectodermal structure developed by the infolding of a vesicle from the surface ectoderm, and within this vesicle layers of lens fibres are developed, the oldest being forced towards the centre and the youngest fibres being found upon the surface. There are many types of opacity which may develop as a result of some interference with a normal development of the lens. Most of these opacities are small and are discovered during the course of a routine clinical examination. It is only a few which lead to visual symptoms and these are sometimes to be traced to a definite upset in maternal health during the pregnancy. One particular condition which should be mentioned is rubella, for it is now known that an attack of German measles during the first three months of a pregnancy conveys a definite risk of the child being born with some congenital defect. This may be in the eye, the ear

For Anatomy of the Eye see Figure 1.

or the heart. Cataract due to rubella is a sporadic disorder, but there are some types of congenital cataract which take a familial form and may occur in several members of the family.

This variety of congenital cataract may be unilateral or bilateral and it is of the type known as *zonular cataract*, because the fibres

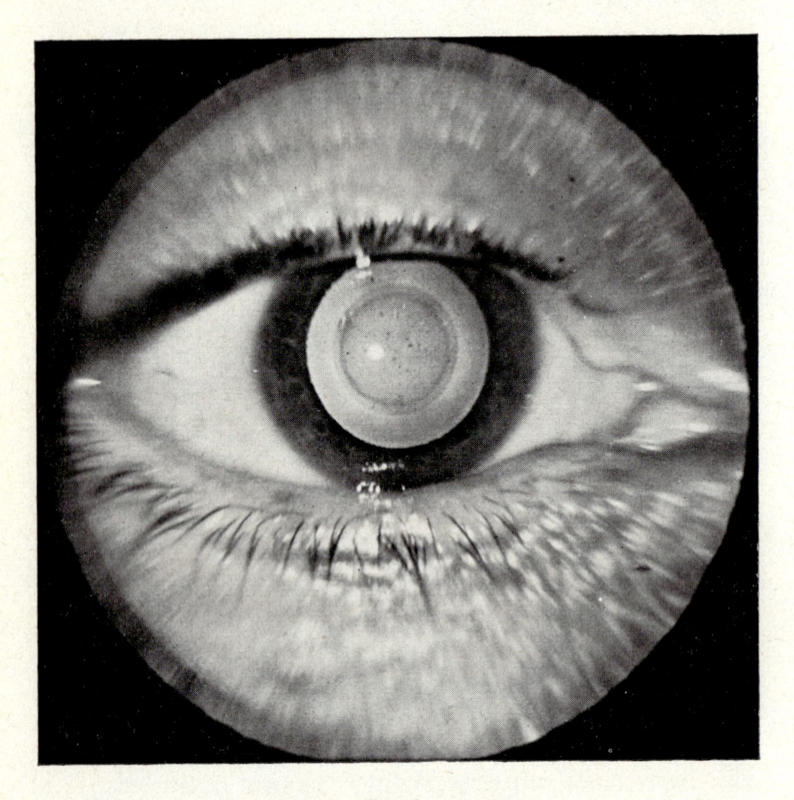

Fig. 27 Zonular cataract.

which are being laid down at the time of the illness are those which are opaque, and one finds clear lens tissue on either side of the opaque band of fibres (Fig. 27).

While the majority of congenital lens opacities do not increase after birth, there are some types, mainly of a punctate variety, in which the opacity becomes increasingly dense in later life and which may then require surgical attention.

Treatment. In the child the need for treatment depends purely upon the density of the opacity, and if the child is able to benefit from normal schooling there is probably no need to interfere, for removal of a lens leads to destruction of the eye's power of accommodation with its consequent disabilities. Operation may sometimes be requested in a unilateral case, if the opacity is so dense as to produce a white appearance in the pupil. This is sometimes sufficiently obvious to be unsightly.

If it is decided that treatment is indicated in a given case, surgical removal of the lens is usually successful and holds out the only hope of producing useful vision. The early onset of visual defect and the presence of an associated nystagmus often make the ultimate visual result disappointing.

Surgical treatment of congenital cataract is by discission, or 'needling'. An incision in the lens capsule allows access of the aqueous humour to the lens material, which swells and is, in the course of time, dissolved by the ocular fluids, or, alternatively, the soft lens matter is aspirated at the time of the initial operation. If complete absorption of the opaque material does not result from an operation, it can be repeated. On clearance of the pupil, the child wears a convex spectacle which replaces the lost refractive power of the lens of the eye, or a contact lens, but restoration of binocular vision is not usually to be looked for in cases of a unilateral congenital cataract treated surgically.

After such operations, delayed complications are not uncommon, particularly detachment of the retina, and it is this danger that forces us to avoid operation for as long as possible. Cataract extraction becomes safer the older the patient.

Senile Cataract

This is the most important group. It accounts for 25 per cent of applications for admission to the Blind Register and is essentially treatable.

Senile cataract is always a bilateral disease, though the opacity in one lens tends to be more advanced than in the other. Treatment is indicated only if the opacity is producing visual defect sufficiently great to interfere seriously with the patient's normal occupation and recreations.

Examination of any individual over the age of sixty-five will reveal some degree of opacity of the lens, and it is only in those cases where serious interference with vision results that treatment is demanded. One should also remember that the patient with senile cataract, if he retains useful vision in the eye, is happier using his one good eye than in having the cataract removed from his worse eye. The reason for this is an optical one, the patient not being able to use together an eye in which the lens is present with one in which the lens is replaced by a spectacle glass.

In an elderly patient is is important to explain why the cataract is not removed from the worse eye while the opposite eye still retains useful vision. Removal of a cataract renders an eye markedly hypermetropic and correction of vision after operation demands the fitting of a strong convex spectacle glass. Even though the corrected visual acuity is good with this spectacle, there is a considerable difference in size between the retinal image received by the aphakic (lens-absent) eye and that received by the normal eye on the other side. This difference in size is in fact of the order of 30 per cent and these two images are so dissimilar that they cannot be fused together as one. The result is intolerable diplopia and is the chief argument against the removal of a cataract in an elderly patient while the opposite eye retains good vision. In younger patients most of the problems connected with unilateral aphakia can be overcome by the fitting of a contact lens which reduces the image size difference to tolerable limits. Contact lenses are being increasingly used after cataract extraction in the elderly. Such a lens improves the field of vision and avoids the need for wearing a rather cumbersome thick spectacle after the operation; and it is expected that their use will become increasingly popular in the more active patient, while the majority of senile aphakics will remain content to use spectacles.

The natural development of senile cataract is towards a gradual increase in the opacity, but it is not possible to say in any given case at what speed the increase will occur. Some cataracts become extremely dense in a period of a few months while others develop slowly and scarcely perceptibly. There is, however, no known medical means of preventing the development of lens opacity. It is as much part of the process of ageing as is the greying of the hair. No drops or other local applications or general treatment are known to have any effect on the cataract. Whatever reputation a medical measure may have depends upon the clinical fact that the vision of a cataract patient varies from day to day and from week to week. What is certain is that the general progress of the condition is towards visual deterioration and that there is no effective method of treatment other than surgery, if accurate correction of the refractive error by spectacles does not produce a useful standard of vision when combined with efforts to obtain the best possible light for reading, a measure which is of great importance in obtaining the best possible function in the presence of lens opacities.

Diagnosis. The first, and often the only, complaint of a patient suffering from senile cataract is visual deterioration. This visual defect is as of a fog or a haze in front of the eyes, usually worse on one side than the other and commonly more obvious in distant

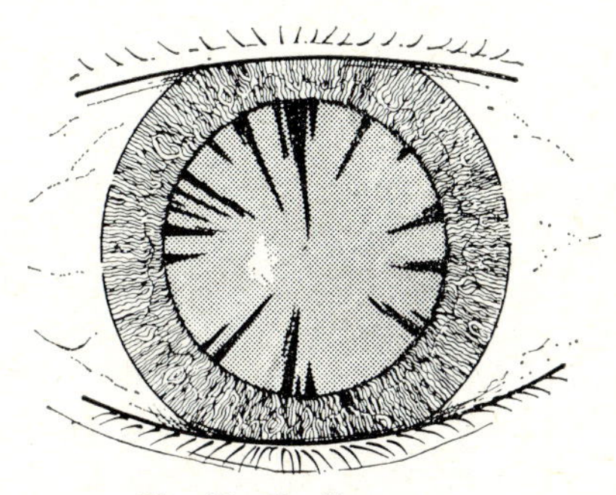

Fig. 28 Senile cataract.

vision, reading being relatively well preserved. Some patients, particularly those with central lens opacities, will say that their vision is significantly worse under conditions of bright illumination when the pupil is naturally small than on a relatively dull day. Examination of the outer eye reveals no abnormality, and ophthalmoscopic examination shows that there is difficulty in visualising the deeper parts of the eye. The red reflex from the pupil is broken up by streaks or spokes of opacity in the case of a common variety of senile cataract with opacities mostly situated in the periphery of the lens (Fig. 28), or there is a central haze interfering with the view with the ophthalmoscope, in which case the opacity is mainly nuclear.

As has been said, surgical treatment of the cataract is indicated when vision in the better eye is sufficiently bad for the patient no longer to enjoy his normal pursuits. It must, however, be remembered that there are occasional cases in which removal of a unilateral senile cataract is recommended. Operation is needed in those cases in which the opacity is so dense as to lead to difficulties due to reduction in the field of vision on the affected side, and those cases in which the cataract threatens to become 'hypermature' with consequent risk of degenerative changes in the eye, particularly iritis and secondary glaucoma.

Treatment. No discussion of surgical technique is proposed, but there are two methods of cataract extraction in current use. These are:

1. The *intracapsular* extraction, in which the lens is entirely removed from the eye complete in its capsule.

2. The *extracapsular* method, in which an incision is made in the lens capsule, the opaque lens matter is removed and the empty capsule remains within the eye.

Each of these techniques gives very good visual results, but the intracapsular method is now more popular and is followed by a shorter convalescence and less post-operative complications.

The days when a cataract operation required of the patient great physical endurance have gone, and it can be said that the period in hospital is likely to be something between one and two weeks, depending upon the views of the surgeon concerned, and some restriction is likely to be placed on the patient's activities during the first few days. The operation may be performed under local anaesthesia with basal sedation, but many surgeons are turning increasingly to general anaesthesia for cataract extraction, and modern techniques allow this to be done safely. A period of some weeks is allowed to pass before the final ordering of glasses for the operated eye, but the patient is usually provided with spectacles at the end of one and a half to two months. The visual results are excellent, and there is no upper age limit to prevent patients from benefiting from cataract surgery. Also, there are virtually no general medical conditions which will prevent a patient from undergoing this operation, which should produce little or no systemic disturbance.

Complicated Cataract

This type of cataract might well be called 'complicating' cataract, for it occurs as a complication of some pre-existing disease within the eye. Among the conditions in which cataract appears as a complication are the following: chronic intraocular inflammation (iridocyclitis, uveitis, choroiditis), chronic retinal disease (e.g. retinitis pigmentosa, total detachment of the retina, myopia), intraocular tumours, and absolute glaucoma.

Treatment. Treatment is of the underlying pathological process, but surgical removal of the lens may be indicated.

Diabetic Cataract

There are two varieties of diabetic cataract, one of which is common and the other rare.

1. TRUE DIABETIC CATARACT. This is a condition usually seen in young patients suffering from severe diabetes, and it is due to sudden interference with intraocular metabolism. The lens changes, therefore, are mostly to be seen in the superficial layers of the lens cortex, where a number of dot-like opacities appear, later becoming confluent. In the early stages these opacities are reversible if the diabetes can be brought under control, but they later become fixed.

2. SENILE CATARACT OCCURRING IN A DIABETIC. Diabetics tend to develop cataract at an earlier age than do non-diabetic patients, and it is, therefore, the practice in eye hospitals for cataract patients to be examined to exclude the presence of diabetes. As in the case of other surgical operations, uncontrolled diabetic patients stand cataract surgery less well than non-diabetics. Apart from the upset which may result from the sudden confinement of a diabetic to bed under a fairly strict regime, it is known that the diabetic patient is more prone to suffer from haemorrhage during and after the operation, and is more liable to develop reactionary inflammation post-operatively. Apart from these considerations, and the fact that surgical results in diabetic patients are sometimes disappointing owing to the fact that removal of the cataract may reveal the presence of diabetic retinal changes, there is no essential difference between the handling of this type of cataract and those in the senile group.

Traumatic Cataract

One of the functions of the lens capsule is, by its osmotic properties, to maintain the correct metabolic conditions in the lens tissue for the preservation of clarity.

If the lens capsule is injured, opacification of the fibres results; this being due to the lens fibres being exposed to the aqueous humour. If the hole in the lens capsule is small, the resultant opacity may also be small and may remain localised, due to the fact that the resultant swelling of the lens fibres blocks the hole in the capsule and prevents further access of aqueous. If, however, the tear in the capsule is large, the opacity of the lens fibres gradually spreads until it involves the entire lens. Not only does opacification occur, but lens swelling takes place and this may be so great that the entire anterior chamber of the eye becomes filled with flocculent lens matter, with subsequent interference with the drainage and circulation of intraocular fluid, and secondary glaucoma.

Cataract may also result from contusion of the eye without rupture, apparently by interference with lens metabolism. It will be seen, therefore, that traumatic cataract may result from a blow on the eye producing no penetration, or from a penetrating injury of the globe, whether this be caused by some sharp object penetrating the coats of the eyeball and then being withdrawn, or by a minute fragment (usually of metal) flying from a hammer or a grinding wheel.

The treatment of such a case depends upon a number of factors, among which are the man's age and the state of the opposite eye, the nature of his employment and the need or otherwise of dealing with the cataract surgically on clinical grounds. By this last consideration is meant the presence or absence of irritation of the eye, or alteration in the intraocular tension, due to the presence of swollen lens matter within the globe. In general, the unilateral traumatic cataract can safely be left untouched, providing the eye settles down comfortably and no metallic foreign body is retained within the eye, for the patient, even were the cataract to be removed, would not be able to use together the two eyes in binocular vision (see Senile Cataract). There are some exceptions to the rule, particularly in the case of the man who requires increased field of vision on his otherwise blind side, and in these cases, surgical extraction of the lens may be called for. The possibility is that the aphakic eye, that is, the eye from which the lens has been removed, may be able to be fitted with a contact lens, in which case restoration of full binocular vision sometimes occurs. Such cases are, however, in the minority and the general practice is, as has been stated, to leave unilateral traumatic cataracts untreated, in the absence of special indications to the contrary.

Other Cataracts

Other possible causes of cataract are:

1. Endocrine cataract, which occurs in tetany, myotonia atrophica and mongolism. There is accumulating evidence that prolonged use of steroid drugs systemically sometimes leads to the development of cataract.
2. Radiation cataract, which occurs as a result of exposure to X-rays, radium and atomic explosions.
3. Heat cataract, which is seen in chain-makers.
4. Metabolic upsets, among which there is galactosuria.

This is by no means a complete list, but will give some indication of the type of condition which may interfere with the clarity of the lens.

CHAPTER EIGHT

Diseases of the Retina

The various vascular diseases of the retina are discussed elsewhere. It is proposed here to describe some of the commoner retinal degenerations and disorders and to discuss their symptomatology.

Senile Macular Degeneration

This is a disappointing and somewhat depressing condition, for it occurs at the age when reading is increasingly important to the patient and it leads to reduction of the central vision. The aetiology is obscure, though there is no doubt that the majority of cases are associated with arteriosclerosis. It is presumably, therefore, a gradual interference with the nutrition of the central area of the fundus which leads to degenerative change. Certain cases, however, are of a hereditary nature, for one sees cases of macular degeneration, occurring in a younger person and indistinguishable clinically from those occurring in the elderly, with a history of similar disease in the family.

The patient's complaint is of distortion or gradually increasing disturbance of central or reading vision. The complaint is often of lack of regularity of lines of print, one letter or word being at a different level from its neighbours. Sometimes, if a scotoma is present, there may be apparently missing letters or words. This is not associated with any failure of the peripheral visual field, and there is at no time any difficulty with getting about. Examination with the ophthalmoscope reveals pigmentary change at the macula, but this change may be of slight degree and it is necessary to dilate the pupil in order to examine the macular area in detail. The ophthalmoscopic appearance of the macula seems to bear little relation to its function, for one sees cases of gross macular disturbance in which visual acuity remains relatively good, and yet there are also patients showing minimal visible disturbance at the macula and gross failure of function. Not only may there be pigmentary changes in cases of macular degeneration, but exudates and colloid-like formations together with retinal haemorrhage also

For Anatomy of the Eye see Figure 1.

occur. The picture is one of extreme variability. No treatment apparently affects the condition, although attention to the patient's general health, particularly the cardiovascular state, is no doubt valuable. It is sometimes possible to improve the vision and to allow reading by various optical devices, particularly magnifying glasses and special telescopic spectacles, and the patient may be assured that this is not a condition that leads to total blindness, and that peripheral vision will be preserved. If the ability to read print is lost, a Talking Book machine is often appreciated.

Retinitis Pigmentosa (Primary Pigmentary Degeneration of the Retina)

This is another hereditary and little explained condition, but one which is not very uncommon. It usually begins to make itself felt in early life, perhaps even while the patient is in his teens and the first complaint will often be of failing vision under conditions of poor illumination. This night-blindness is characteristic of retinitis pigmentosa and is associated with a gradual increase of retinal disturbance, manifest by increasing failure of vision. The ophthalmoscope shows that there is disturbance of retinal pigment, in the form of deposits of superficial retinal pigmentation, mostly seen in the mid-zone of the fundus. Clumps and aggregations of branched masses of pigment are to be seen and there is often a ring-like disturbance of the visual field, fitting in with the distribution of the retinal degeneration. As time goes on (and it is impossible to predict the rate at which any particular case will advance) the retinal pigmentation becomes of greater degree and involves an increasing area of the fundus; it is associated with pallor of the optic nerve head and constriction of the retinal blood vessels, and with the development of lens opacities. This complicated cataract increases the patient's visual handicap. No treatment is of avail. The visual outlook is grave and many of these patients become grossly handicapped in later life.

Detachment of the Retina

The retina may become raised from its bed as a result of a little-understood process which we call *simple* or *idiopathic detachment of the retina*, or it may be raised by the presence of some inflammatory or neoplastic disease. These two types of retinal detachment are important in that it is necessary for them to be distinguished clinically. The handling of the one is entirely different from that of the other.

SIMPLE DETACHMENT OF THE RETINA. This is liable to occur in myopes, for the short-sighted eye is often larger than the normal eye and the retina degenerative, particularly in myopia of moderate degree. Simple detachment of the retina may also occur after an injury, and in this case it is the contusion of the globe which produces a tear in the retina and results in the detachment. Retinal detachment is one of the possible delayed complications of intra-ocular surgery, especially cataract surgery.

The understanding of the pathological processes involved is assisted by the knowledge that the retina developmentally consists of two layers which become apposed to one another and which remain closely in apposition though not physically bound together. If a collection of fluid obtains access to this potential space then the inner or neural layer of the retina is displaced towards the centre of the eyeball.

This process is what occurs in simple detachment of the retina. A hole or tear appears at some part of the retina and through this hole fluid from the vitreous gains access to the space between the two retinal layers. The resultant displacement leads to failure of vision, and the patient will complain of a gradual painless visual loss which may have been ushered in by a sensation as of sparks or flashing light, a phenomenon related to the original occurrence of the actual retinal tear. A thoughtful patient will often describe the area of his visual field in which the retinal detachment occurs, and he will say that it is as of a curtain coming across his field of vision from one side or another. This phenomenon is due to the fact that detachment of the retina commences in one region of the retina and spreads gradually (Fig. 29).

Examination through the dilated pupil shows that there is, in part of the fundus of the eye, a loss of the normal red reflex from the retina, and that the retinal vessels where they pass over the billowing surface of the detached area are not as straight and well defined as they are in other areas. The vessels in fact become more sinuous and convoluted and appreciably darker in colour. Examination of the visual field shows field loss corresponding to the area of the detachment.

Treatment of a retinal detachment is purely a surgical problem. An attempt is made to find and seal the hole in the retina through which the fluid is entering the intraretinal space. If this can be done and the fluid between the two layers of the retina evacuated then there is a good prospect, in a recent detachment, of surgical success. In early cases, showing a well-defined hole and an otherwise healthy eye, the success rate is somewhere in the region of 80 to 90 per cent. Although the use of a diathermy current to

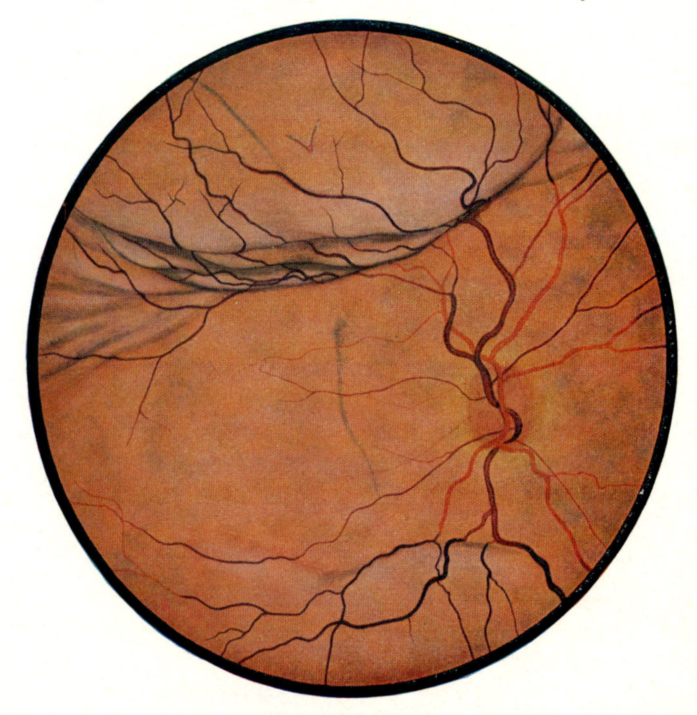

Fig. 29 Simple detachment of retina. ('Arrow-head' tear above.)

produce an inflammatory reaction and subsequent adhesion between retina and choroid has been in use for many years, this method is now being replaced by newer techniques. Refinements in retinal detachment surgery have led to plastic procedures on the sclera to produce infolding of the outer coats of the eye (scleral resection) and to the use of intense cold (cryosurgery) and bright light (photocoagulation) to induce choroido-retinal adhesion. Each case requires individual study and there is no single procedure that is applicable to every retinal detachment.

Retinal detachment is a well-recognised complication of cataract surgery. The reason for this is not entirely clear, but it is probably connected with the lack of normal support for the retina due to the removal of part of the ocular contents and the extreme lowering of the intraocular tension together with the necessary surgical manoeuvres which occur at the time of the operation. The condition is particularly common as a late result of surgery for congenital cataract.

SECONDARY, OR SOLID DETACHMENT OF THE RETINA. A clinical history and appearance similar to that which has been described above may arise as a result of the elevation of the retina from its bed by a malignant neoplasm. This usually arises from the choroid and the two common types of tumour are the malignant melanoma of the choroid and the metastatic tumour, a secondary perhaps from breast or prostate (Fig. 30).

The patient's complaint in the presence of an intraocular neoplasm is the same as in the case of a simple retinal detachment, that is to say, of gradual painless visual loss often of a sectorial nature. If, however, the tumour is large or is situated far forward, secondary glaucoma may arise due to interference with the circulation of the intraocular fluids. This glaucoma may bring the patient for attention on account of the pain and redness of the eye to which it gives rise. In this event the condition has to be distinguished from that of simple congestive glaucoma, or a secondary glaucoma of inflammatory origin.

The distinction with the ophthalmoscope between a simple detachment of the retina and a solid detachment of the retina may be difficult. The solid detachment tends to be globular in nature and often shows new vessel formation on its surface. The picture is somewhat confused by the fact that a solid detachment is not infrequently associated with an exudation of fluid which gives rise to a coincident serous detachment of the retina, often below. Another diagnostic method is to transilluminate the eye with a small bright source of light. After amethocaine has been placed in the conjunctival sac, a bright light is held against the surface of the sclera in the various quadrants 10 to 15 mm behind the limbus of the cornea. If the light passes freely into the interior of the eye, the globe becomes full of light and the pupil glows. If a tumour is present this transillumination does not occur, and there will be found one or more areas where application of light to the surface of the sclera does not lead to illumination of the pupil.

In general, treatment of intraocular malignant neoplasms is by excision of the eye. There are, however, certain possible exceptions to this general rule and one is the presence of a malignant neoplasm in an only remaining eye, or in an only useful eye. Under these conditions, radiotherapy may be considered.

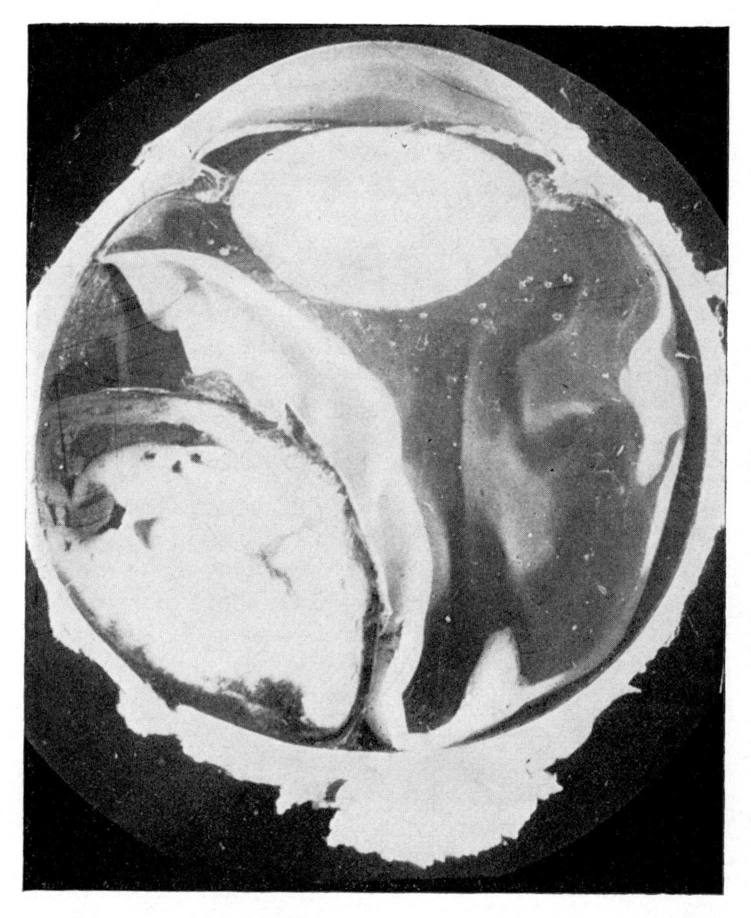

Fig. 30 Tumour of choroid.

Another case in which enucleation of an eye containing a malignant tumour would perhaps not be justified is in the case of a secondary choroidal neoplasm in which the primary growth is irremovable or has disseminated widely. To subject the patient to an operation to remove an eye would probably not be indicated unless the eye became painful. In this case radiotherapy to the eye will sometimes cause regression of the local tumour and help to avoid an enucleation under these distressing conditions.

Retinoblastoma

This is a true retinal tumour and occurs in childhood. The majority of cases are reported in the first three years of life and most are seen during the first year. The condition is often hereditary.

The only complaint that the parents have is that they notice an abnormal appearance in the pupil. This may only be seen when the pupil is partly dilated in a dim light and consists of a whitish reflex from the pupil which appears white instead of dark, the amaurotic cat's eye reflex. This appearance is due to the presence, behind the lens, of tumour tissue and indicates an advanced stage in the condition. Cases are not often noticed at an earlier stage than this, except coincidentally, for it is difficult or impossible to assess the visual acuity of an extremely young baby.

The condition shows a strong tendency to bilaterality and about a third of the cases are bilateral. The second tumour is not a metastasis, but it is a primary neoplasm.

Treatment is difficult and often disappointing. If the tumour in the first eye is of sufficient size to have virtually destroyed the vision, then enucleation of the eye is probably the treatment of choice. If we are faced with the treatment of a tumour in the second eye in the same child, then radiotherapy offers a possible means of attack, for these tumours are usually radiosensitive. The application may be of actual X-rays, or of radioactive material to the outside of the sclera over the growth by fixing radon seeds or other material in a plastic applicator. Response to radiation is often good and shrinkage of the tumour occurs. Frequent observation is required thereafter in order to watch for any signs of recurrence and the interval between the observations is gradually lengthened as the child grows up. It is then hoped that the scarring produced by the tumour and by the radiation has left sufficient actively functioning retina to be useful.

If the condition is untreated, or if treatment is unsuccessful, the tumour spreads forwards, producing proptosis and a fungating mass between the eyelids, or it spreads backwards along the optic nerve, enlarging the optic foramen and producing death by intracranial disturbance. These tumours also show a tendency to metastasise and spread to lung and bones in particular.

One point of diagnostic importance which might be mentioned here is that these tumours not infrequently show calcification, and X-ray of the orbit may be helpful in the diagnosis.

Eclipse Burn of Macula

The unguarded observation of an eclipse of the sun can lead to a burn close to the macula. A central visual defect is at once noticed and this scotoma decreases in size during the subsequent weeks. Some permanent defect may remain and a small pigmented scar will be seen with the ophthalmoscope.

CHAPTER NINE

Errors of Refraction and Ocular Headache

The similarity between the eye and a simple camera has been repeatedly described and, as long as it is not carried too far, the analogy is helpful.

When the eye regards an object, an inverted image falls on the retina. This image gives rise to photochemical changes in the receptor cells, and the stimulus is converted into an electrical impulse which, in turn, passes to the cerebral cortex, where it is analysed.

The reception of a clearly focused image depends on the presence of a normal relationship between the length of the eye and the focal length of the 'lens system'. The expression 'lens system' is used to describe the focusing mechanism of the eye, for there are two factors involved in the refraction of the rays of light entering the eye. One is the biconvex lens itself, which is suspended immediately behind the iris, and the other (more powerful from the optical point of view) is the smooth convex anterior surface of the cornea (Fig. 1).

Accommodation

If the eye is to produce a clear retinal image of objects at varying distances, there must be a mechanism which will adapt the focal length of the lens system to suit the varying nature of the entering rays. This is achieved by alteration in the curvature of the lens of the eye.

Consisting of a transparent mass of lens matter, enclosed in an elastic membrane (*the lens capsule*), the lens is suspended within the circle of the ciliary body by a series of fine fibres (*the zonule or suspensory ligament*). The ciliary body contains a mass of muscle fibres, innervated by the parasympathetic element of the third nerve, and this muscle by its activity is able to alter the tension existing in the suspensory ligament and thus in the lens capsule. Contraction of the ciliary muscle leads to an increase in the

For Anatomy of the Eye see Figure 1.

curvature of the lens and thus to an increase in its effective power. This produces focusing of the eye for close objects (Fig. 31).

The stimulus for this movement of accommodation is a reflex one, arising from the need to maintain a clear retinal image. It should also be said that there is a natural relationship between the amount of accommodative effort required to produce a clear

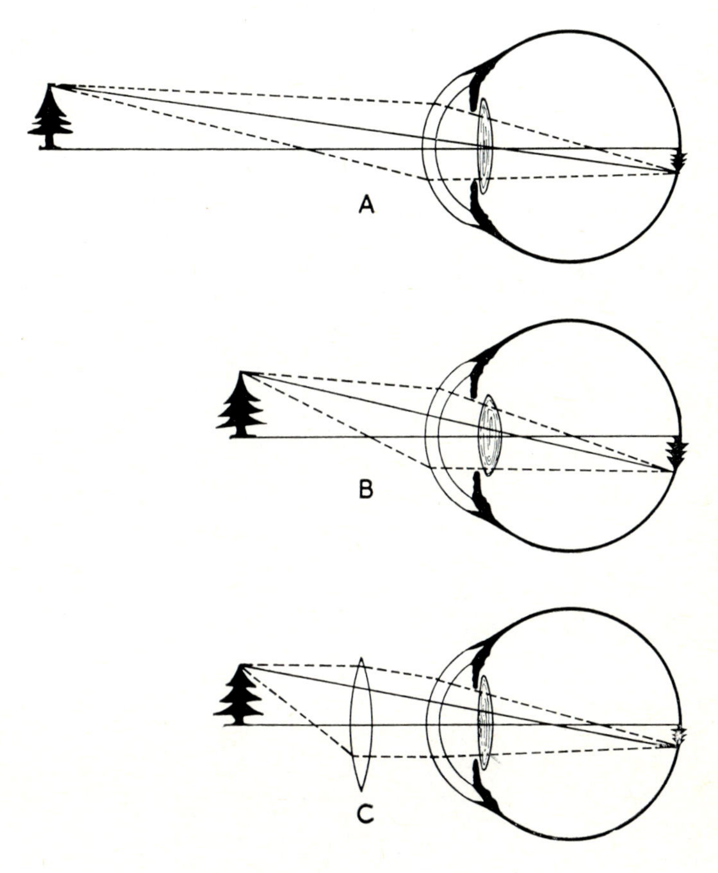

Fig. 31 Formation of the retinal image.

(A) In the normal eye, with the accommodation relaxed, the image of a distant object falls on the retina. (B) When the object approaches the eye a change of shape of the lens takes place and the image remains focused on the retina. This is accommodation. (C) With increasing age the eye loses its power to change its focus for close objects and has to be reinforced by convex lenses to keep the image in focus. This is presbyopia.

retinal image of a close object and the amount of convergence of the visual axes needed to keep this image on the maculae of the two eyes. The importance of this relationship and of the factors which lead to its disturbance are considered in the description of refractive errors in their relationship to the aetiology of squint.

Presbyopia

As age increases, the lens of the eye becomes increasingly sclerosed and less able to respond to the efforts of the ciliary body to produce changes in its curvature. Clinically, this is manifest, at about the age of forty-five, by increasing difficulty in reading small print and the need to hold reading matter at an increasing distance from the eyes. This phenomenon is seen at an earlier age in hypermetropes than in myopes and is gradually progressive until the age of about sixty-five, when all power of accommodation is lost.

The development of presbyopia depends purely on the optical condition of the eyes and is uninfluenced by treatment. There is no evidence that the wearing of spectacles, either before the onset of presbyopia or after it, has any effect on its development.

Treatment of presbyopia involves the provision of convex lenses of suitable strength to allow work at the normal reading distance. One of the disadvantages of reading glasses is that clear distance vision is not possible in them, and it is often necessary to provide bifocal glasses, or 'half-moons' to overcome this.

The Aetiology of Refractive Errors

The classical description of hypermetropia and myopia includes the statement that the eye is either 'too short' or 'too long' and that, therefore, the image formed by the lens system falls either behind, or in front of, the retina. It is admitted that in some cases, particularly the higher degrees of error, the actual physical length of the eye plays a part in the causation of the refractive state, but the classical description requires some modification in the light of recent knowledge. It is now known that the normal state of affairs (*emmetropia*) is the product of an accurate balance between the length of the eye and the focal length of the refracting media. It is when this relationship is disturbed that refractive errors arise and we now know that both hypermetropia and myopia may occur in eyes of average length.

Hypermetropia (Long Sight)

This is the commonest refractive error. It is practically universal in infants and becomes gradually less as age increases.

In hypermetropia, rays of light from an object in the distance are brought to a focus (if accommodation of the eye is relaxed) behind the retina. If the individual is young, and the degree of

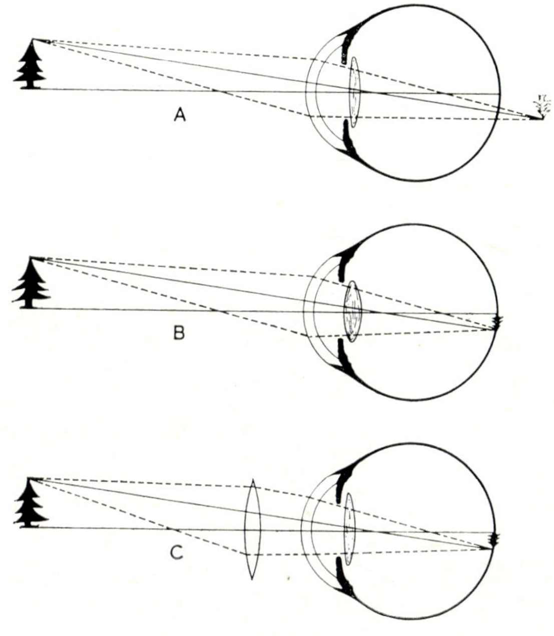

Fig. 32 Hypermetropia.

(A) With the accommodation relaxed, the image of a distant object falls behind the retina. (B) An effort of accommodation is required for the clear viewing of an object, even in the distance. (C) This effort has to be increased as the object becomes closer. If the error cannot comfortably be overcome by accommodation, a convex spectacle will be needed.

hypermetropia relatively slight, this state of affairs is no disadvantage, for it can be compensated for by contraction of the ciliary muscle, the effect of which is to increase the refractive power of the lens and so to bring the image forward on to the retina. In higher degrees of hypermetropia no disadvantage may be felt in distant vision, but the increased accommodative effort required to focus near objects may give rise to symptoms of eyestrain (Fig. 32).

This explains why hypermetropes require glasses for reading at an earlier age than usual for, with the gradual failure of accommodation that occurs with increasing age, vision at close range is the first to suffer.

Hypermetropia, therefore, demands correction only if it is of such a high degree as not to be overcome by the use of the eye's own focusing mechanism, or if it is giving rise to symptoms of eystrain. (The exception to this rule is that of the child suffering from convergent squint, in whom the refractive error often has to be fully corrected as an essential part of the treatment of the condition.) There is certainly not the slightest justification for the view that every case of hypermetropia demands optical correction.

It will be appreciated that a degree of hypermetropia, insufficient to cause symptoms in youth, may lead to difficulty in middle-age when the natural lessening of accommodative power takes place. There will be an earlier demand for reading glasses than in the case of the patient with no refractive error (the emmetrope).

The correction of simple hypermetropia involves the provision of convex spectacles of such a strength as to allow relaxation of the patient's accommodation in the viewing of distant objects. Then the inspection of near objects can be carried out with the exercise of the normal amount of accommodation.

The hypermetrope of moderate degree will at first need to wear his glasses for close work. With increasing age and the natural diminution in the power of accommodation, a correction will also be required to allow clear distance vision.

Myopia (Short Sight)

In myopia, rays of light from a distant object are focused in front of the retina and the myope is, therefore, at a disadvantage when compared with a hypermetrope in that he is unable to obtain clear distant vision by the exercise of his accommodation. In fact, accommodation in myopia will only worsen the vision in the distance, as an increase in the refractive power of the lens increases the blurring of the image (Fig. 33).

At close distances, on the other hand, the advantage is with the myope, who is able to obtain clear vision for near objects with little or no effort of accommodation. This accounts for the tendency for the myopic child to prefer books to outside games, and often to be rather shut off from his fellows. In addition, the abnormally close reading position which myopia promotes leads to round shoulders and a poor general posture. In seeking employment the myope takes to trades in which the ability to do fine close work with little effort is an advantage.

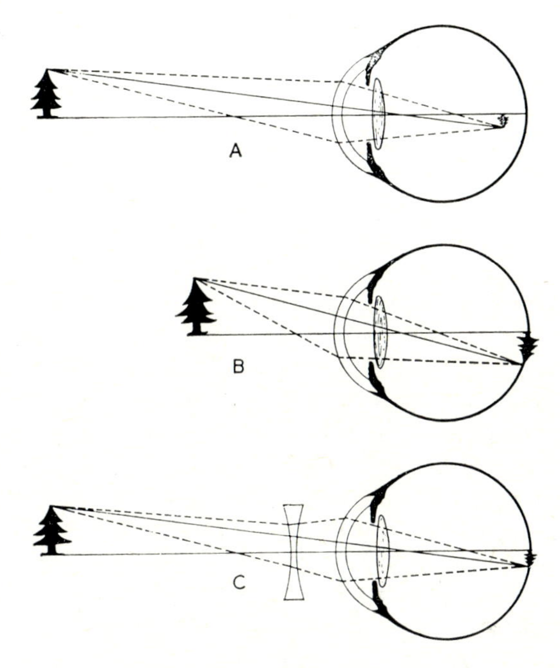

Fig. 33 Myopia.

(A) The image of a distant object falls in front of the retina and any effort of accommodation will only increase the blurring. (B) Near objects, on the other hand, are seen clearly with little or no accommodation. (C) For clear distant sight, concave glasses are needed.

Myopia is strongly hereditary, and the lower degrees of error appear generally as a dominant characteristic. Happily, the same does not apparently apply to the more destructive and incapacitating high degrees.

Myopia is of two types:

1. *Simple myopia*, which is common.
2. *Progressive*, or *degenerative myopia*, which is less common.

SIMPLE MYOPIA. Although this condition may reach a fairly high degree, it does not show the secondary changes which characterise the second variety. The onset is usually during the school years (hence the term 'school myopia'), but it may be delayed into early adult life.

The present system of routine examination of schoolchildren usually leads to the early recognition of the state of affairs, or the child's teacher notices that the blackboard is not clearly seen from the back of the class. In pre-school children, the parents may notice that reading matter is held abnormally close to the eyes; though this is not an uncommon finding in children with perfectly normal eyes who are just starting to read. In the latter case, it is a manifestation of psychological, rather than visual, difficulty.

Simple myopia advances during the growing years and then becomes stabilised. The ultimate degree of error cannot be predicted in a given case. Yearly or half-yearly examinations are necessary at first and concave lenses are prescribed to give the child full distance vision. He is then able to lead a normal life, can play all games (unless the error is so high that he is literally incapacitated when not wearing his spectacles) and is on equal terms with his school-fellows. For games, the provision of splinter-proof glasses is a wise precaution. Until recently it was taught that the myopic child should be forced to wear his spectacles from morning to night, but this is considered to be too severe an imposition to place on a youngster who probably will not take kindly to wearing glasses in any event. Providing the child wears his spectacles all the time while in class, and at other times when clear distant vision is required, he can usually be allowed to decide for himself, as he grows older, whether or not he will wear his glasses for everyday life.

The brothers and sisters of myopic children should be examined during their early schooldays to exclude a similar condition and any deterioration in distance vision should be watched for.

DEGENERATIVE MYOPIA. The aetiology of this condition is obscure. The myopia is of high degree and is associated with pathological changes in the eye. The most important of these changes are:

1. Vitreous opacities.
2. Cataract formation.
3. Degenerative changes in the choroid and retina at the

posterior pole of the eye. This may spread to the macular area and give rise to a gross defect of central vision. Myopes are also particularly prone to detachment of the retina.

The Prevention of Myopia. As the refractive condition of the eye is governed by hereditary and constitutional factors, there is no way of influencing its development. No exercises, diet, medicines, or restrictions on the use of the eyes can affect the process. The only possible treatment is the provision of correct glasses and the maintenance of a good state of bodily health.

Astigmatism

If the front surface of the cornea is a smooth, regular curve, astigmatism is absent, but this ideal condition is not often met. The more usual finding is for the curvature in one meridian to be greater than that in another, with the result that entering rays of light are not regularly refracted and are deviated more in one plane than in the other. This may be associated with either hypermetropia or myopia and cannot be corrected by the simple use of spherical lenses which bend rays of light regularly in all meridians. Its correction requires special lenses which are so ground as to bend the rays in one meridian alone. Such a cylindrical lens, combined with a spherical lens if need be, neutralises the unequal curvature of the cornea.

The majority of cases of astigmatism are due to imperfectly understood influences acting on the development of the corneal curvature but some are due to injury or disease. Among the possible causes are corneal wounds and operation scars, keratitis, and various degenerative processes, especially conical cornea. In certain cases adequate visual improvement cannot be obtained by the use of spectacle lenses, and contact lenses, which act by neutralising the irregularity of the corneal surface, may be indicated.

Ocular Headache

It is often very difficult to decide whether or not a given headache is of ocular origin, for headache is one of the commonest symptoms of which patients complain. It is certainly very prominent in the list of reasons for which patients are referred for ophthalmic examination.

The possible causes of headache are many, and range from

constipation to cerebral tumours. It is not proposed to discuss the diagnosis of headache in general, but only to indicate features which may point to the eyes being implicated in a particular case.

Probably the main point is that ocular headache is usually related to, or precipitated by, the use of the eyes. Thus it occurs on prolonged reading or sewing, in the cinema, and on taking up a clerical occupation for the first time. The pain may be delayed in onset, appearing in the morning after a period of intense ocular activity the preceding evening. It is usually relieved by ocular rest, weekends, summer holidays, and so on.

The site of the pain is varied. It may be felt in the eyes themselves, in the temple, or in the occipital region, in which case it probably arises from the neck muscles. The headache is usually regular in occurrence. A pain appearing at long intervals, without any change in the patient's ocular habits, is not likely to be of ocular origin. Similarly, a headache which is of recent onset, without any change in the use of the eyes, is not likely to be ocular.

Ocular headache, or eyestrain, is sometimes due to uncorrected refractive errors and may occur in the presence of good visual acuity. It is also not uncommonly seen in the presence of negligible, or adequately corrected, refractive errors. In these cases the symptoms may be due to a lack of balance in the oculo-motor apparatus of the two eyes (see Heterophoria).

Patients with migraine sometimes complain of ophthalmic symptoms—flickering lights, hemianopia, etc.—but the headache which follows is not usually influenced by the wearing of appropriate spectacles.

CONTACT LENSES

Lenses which are worn inside the lids and in close contact with the eye owe their value to the fact that they remove the irregular refraction of the rays of light at the surface of the cornea and replace this surface by a smoothly polished surface of glass or plastic.

Cases for which these lenses may be considered suitable fall into the following groups:

1. *Optical.* Those cases in which adequate vision is not obtainable with spectacles. Under this heading come cases of high

myopic astigmatism, conical cornea, some cases of scarred cornea due to past inflammation or wounds, and irregular refraction after corneal graft operations; also aphakia, following cataract extraction.

2. *Therapeutic.* These are cases in which the lens is used to protect the eye, or to prevent the frequent breaking down of degenerative conditions. Among the indications for this type of treatment are mustard gas keratitis and the keratitis of acne rosacea. In some cases not only will the cornea be protected, but the patient will also have improved vision.

3. *Occupational.* In a variety of occupations the wearing of ordinary spectacles may not be practicable; they include those who work in steamy atmospheres where the misting of glasses makes vision difficult, actors and actresses who cannot perform in spectacles, and players of professional games.

4. *Cosmetic.* In these cases contact lenses may be considered simply because the patient is unwilling to wear glasses.

Although contact lenses are theoretically the answer to many optical problems, there are still a number of difficulties to be overcome.

The first of these is that it is impossible to make a prior assessment of the patient's tolerance and he may find himself unable to wear the lenses for any length of time. This disadvantage is becoming less as fitting technique improves.

The fitting of contact lenses is still a tedious process in many cases, and some patients never develop the ability to insert and remove the lenses easily.

Expense is still a major factor. It is only in those cases where the lens is certified as being necessary on medical grounds that these appliances are available through the National Health Service.

On the whole it can be said that only those patients who obtain real visual benefit from their use, or whose comfort is materially increased, should be advised to wear contact lenses. In the future, however, and with increasing experience in the design and fitting of these lenses, it seems likely that a wider field for them will open out. It is already possible to promise a greater prospect of tolerance and wearing time to the majority of patients, and the development of purely corneal, or microlenses, has reduced the number of fittings needed for the production of lenses which can be worn for a useful number of hours, if not all day long. These developments are leading toward the use of contact lenses by an increasing number of young myopes, particularly girls, to whom the constant wearing of spectacles is unacceptable. In many such cases it is coming to be agreed that these lenses are justifiable.

The most exciting future prospect may well lie in the perfection of soft, or hydrophylic lenses. These mould themselves closely to the cornea and the problems of fitting are less than in the case of rigid lenses. Owing to the ability of the soft lens to absorb fluid, however, the maintenance of sterility is of particular importance, and it is not yet clear what effect the soft lens, worn over a long period, may have on corneal metabolism.

SPOTS BEFORE THE EYES

Patients frequently complain of floating opacities in front of the eyes, usually of a 'cobweb' or 'fly' (hence 'muscae volitantes'), and it may be difficult to know what significance to attach to them. Certainly, they are *not* due to the liver and are *not* best treated by purgatives.

All floating spots which can be seen by the patient are due to opacities in the vitreous, usually in a somewhat degenerate and partially fluid vitreous, and they are often stirred up by sudden movements of the head or eyes, returning to rest when the eyes are still. As the vitreous is commonly fluid in myopia, most of these patients are short-sighted. In them, floating specks can be disregarded, and it is not unusual for these to become less obvious as time goes on. Whether this is due to the opacities breaking-up into smaller fragments, or to the patient learning to disregard them, is uncertain. Floating opacities of a similar type occur in the elderly, on account of degenerative changes in the senile vitreous.

There are some rare cases in which floating spots are of more significance and these are cases in which they increase appreciably in a short time and are accompanied by failure of vision. Then they may be more serious, perhaps because they are of inflammatory origin, as in uveitis, or because they are due to intraocular disturbance, such as retinal detachment.

In discussing, with the patient, the nature of his complaint, it is necessary to discover the form of the symptoms. The simple myopic or senile vitreous floater usually has a shape which the patient describes and it is definitely 'floating'. One must remember that some patients refer to a 'spot in front of the eye' when they actually mean that a part of the field of vision is distorted. Such a symptom occurs in retinal vascular occlusion (especially where a single branch of the vein is involved), in detachment of the retina, and in some lesions of the macula. Among the latter are the results of some contusion injuries, early neoplasms at the posterior pole of the eye, and the effects of burns at the macula, as occurs during the unguarded observation of an eclipse of the sun. All these conditions produce the subjective complaint of a 'spot', or defect, bearing a constant relationship to the object looked at, and not floating from one point to another.

CHAPTER TEN

Squint

CLASSIFICATION

1. Non-paralytic squint.

The common squint of childhood. Angle of squint does not vary with the direction of the gaze. (Concomitance.)

2. Paralytic squint.

Either congenital, or acquired in later life following trauma, or some lesion affecting the motor pathways. Angle of squint varies with the direction of the gaze.

3. Latent squint. (Heterophoria.)

NON-PARALYTIC (CONCOMITANT) SQUINT

The development of binocular function in an infant can be watched by any careful observer. At first, in early infancy, the movements of the two eyes often bear little relationship to one another. Particularly is this so when the baby is showing some trivial physiological upset; hunger, wind, anger, and so on. As time goes on, the binocular reflexes become more firmly established and the retina becomes fully differentiated, with the result that the stimulus of a clear retinal image falling on the two eyes produces a reflex movement of the eyes to bring the image on to the most sensitive retinal area, the macula. It is in fact true to say that the child has to learn to use the two eyes together in the act of seeing in the same way as that in which it has to learn to use the two legs together in the act of walking. It is equally true that the reflex act of seeing is just as easily upset by an obstacle acting during the early period of its development as is the act of walking, which has only recently been learned.

The most common obstacles leading to the failure of the development of binocular vision or to its breakdown in the early years of life are:

1. High degrees of refractive error.

For Anatomy of the Eye see Figure 1.

2. Ocular disease.

3. Failure of the proper development of the cerebral mechanism concerned with the appreciation of binocular vision.

ERRORS OF REFRACTION AND SQUINT

The relationship between these two is so important that there are some who hold that all children should be examined at the age of two or three years and that significant errors of refraction should be corrected to prevent the later development of squint. This solution is impracticable at present, but there is no doubt that the problem would be very much reduced and the duration of active treatment lessened if every squinting child were to be provided with spectacles, if necessary, to correct his refractive error, *immediately on the squint appearing for the first time.* For a proper understanding of the position, some description of the way in which these squints arise is necessary. This can be given without much use of technicalities.

A hypermetropic, or long-sighted, individual cannot obtain clear vision, even in the distance, without exercising his accommodation to an extent which depends on the degree of his hypermetropia. In youth, the ciliary muscle is powerful and the lens malleable, so quite appreciable degrees of hypermetropia can be overcome in this way. For the examination of close objects, as in reading, the eye requires to be focused for the near distance, thus adding to the accommodative effort already being exerted by the ciliary muscle.

Under normal conditions there is a balanced relationship between the effort being exerted by the ciliary muscle, in the act of accommodation, and that being put out by the internal rectus muscles, in the act of convergence; the two processes being coordinated to bring a clearly focused image on to the macula of each eye, at whatever distance the object (Fig. 34).

If now, as in the hypermetropic child, the accommodative effort is out of proportion to the degree of convergence required and, as has been said, the reflexes concerned in the maintenance of binocular fixation have only recently been developed, the stimulus to convergence may be so great as not to be resisted, and overconvergence occurs. The child then adopts uniocular vision and allows one eye to turn in.

It is in this way that the majority of the squints with which we are concerned develop.

Ocular Disease in the Causation of Squints

If there is any obstacle preventing the proper reception and perception of the retinal image, the stimulus of macular fixation is lacking and an eye may squint. There are many possible causes

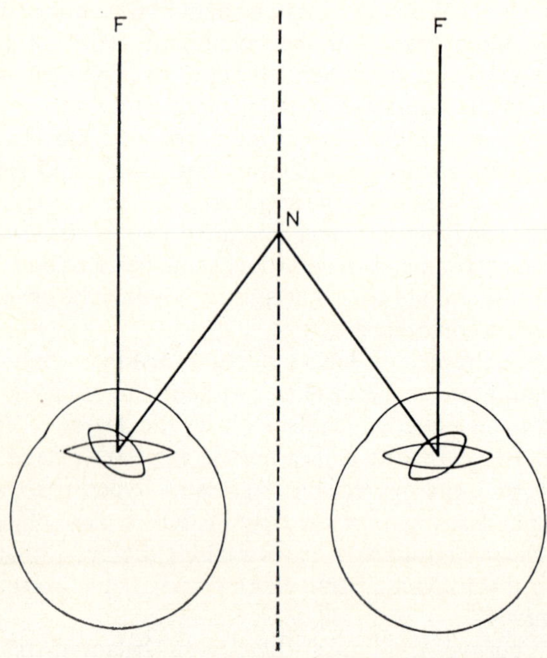

FIG. 34 The Accommodation–Convergence Relationship. If the eyes are considered, in the first instance, as regarding a distant object (F) a clear retinal image is obtained in each eye without any effort of either accommodation or convergence. On looking at a near object (N) a balanced relationship normally exists between the degree of accommodation exerted and the angle of convergence. In hypermetropia this relationship is upset.

of such failure and an essential part of the investigation of any squint is an examination to exclude such a cause. Among the conditions which have to be considered here are corneal opacities, such as may follow trauma or keratitis, congenital cataract, disease of the retina, particularly intraocular tumours and macular disease, and optic atrophy.

Failure of the Development of the Cerebral Mechanism

Some squints develop in the absence of significant refractive error. They usually appear early and are alternating in character. That is to say, the child uses either eye with equal facility, and changes fixation from one to the other. To explain this phenomenon, there must be postulated some defect in the mechanism of perception, presumably of developmental origin.

Clinical Varieties of Non-Paralytic Squint

Convergent or divergent.
Constant or intermittent.
Uniocular or alternating.

History of a Typical Accommodative Squint in a Hypermetropic Child

AGE OF ONSET. Three to six years. At this age the visual acuity is almost fully developed and the child is either going to school for the first time or is beginning to take an interest in books and small pictures. It is surprising how often a squint is first noticed by someone other than the parents; a neighbour, grandparent, or the family doctor.

TYPE OF ONSET. Probably all these squints are intermittent at first, most often occurring when the child is tired, angry or unwell, or at mealtimes. A general illness, such as measles or whooping cough, will sometimes precipitate the onset.

SUBSEQUENT PROGRESS. *A child does not 'grow out of' a squint.* The deformity becomes more frequent until it is constantly present. Time spent in waiting for natural cure to take place is time wasted. The degree of squint (the size of the angle between the axes of the two eyes) is of no importance in this connection. A small angle of squint demands attention just as urgently as does a large one.

The child, now using only one eye for vision, neglects or disregards the image received by the squinting eye, in order, originally, to avoid diplopia (Fig. 35 and 36). Due to this disuse, the visual acuity of the squinting eye falls. It seems that the central mechanisms concerned with the reception of the visual impulse 'get out of practice' and this leads to a gradual loss of sensitivity. The condition so developed is *amblyopia* ('lazy eye') and is the most important single complication of a squint, to the handling of which most of our energies have to be directed in the early stages of treatment. The importance of amblyopia lies in the fact that, if it is allowed to persist, it becomes permanent, and no treatment is then effective in restoring vision to the affected eye. The time that this permanent amblyopia takes to develop is variable, depending largely on the age of the child at onset and, therefore,

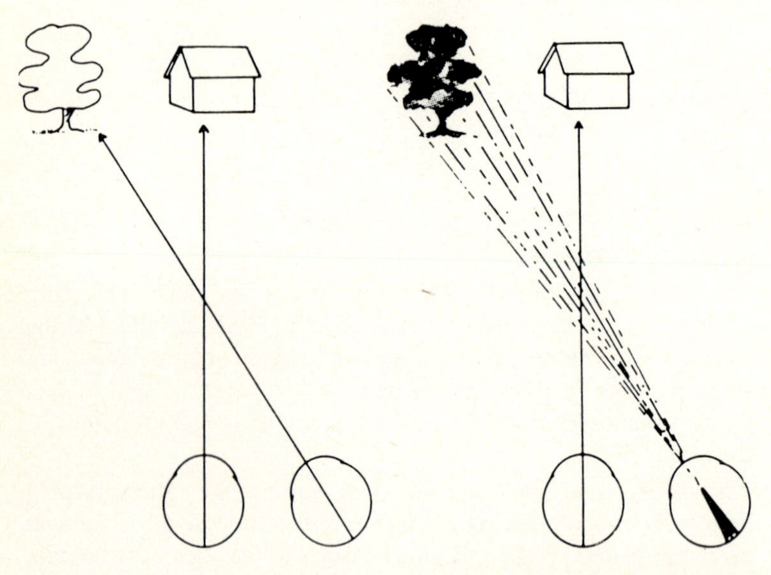

Fig. 35 Double vision in squint.

Fig. 36 Suppression of image from squinting eye.

on the degree to which vision had developed before the squint appeared, but, as a general rule, it may be said that energetic treatment applied before the age of seven years may well be effective in curing amblyopia. After this age the outlook is far less certain.

Diagnosis

1. APPARENT SQUINT. There are certain conditions which give rise to the appearance of a squint; though the visual axes are actually parallel. The commonest cause of this phenomenon is *epicanthus* (Fig. 12), when the bridge of the nose is very broad and an abnormal fold of skin hides the medial angle of the eye.

The corneae then seem to be closer to the midline than they actually are and convergent squint is mistakenly thought to be present. The true state of affairs will be disclosed by the application of the *cover test* (see below).

2. TRUE SQUINT. In large degrees of deformity, and particularly in those squints which are uniocular, there is usually no doubt as to the diagnosis. It is in the case of smaller angles that doubt will often arise. The cover test is designed to expose a squint of small degree.

The Application of the Cover Test

1. The child is persuaded to fixate an object—a light, a doll, or a brightly coloured toy.

2. While fixation is being maintained, the eyes are covered in turn, either by the hand or a card. If each eye remains stationary when its fellow eye is covered, no squint is present. If, however, on covering one eye the opposite eye moves in order to take up fixation, this eye was originally squinting. When the cover is removed the originally squinting eye may maintain fixation; when it will be found that the covered eye is now squinting (*alternating squint*), or it will revert to its original position and the eye which has been covered will take up fixation once more (*uniocular squint*). In older children, the cover test should be carried out while the child is first looking at a distant object and then at something held a foot or so from the eyes. This additional measure may disclose a squint which is not present at all distances of the gaze.

TREATMENT

This follows naturally on the principles of aetiology which have been outlined and consists of the following stages, though not all of these processes may be needed in the treatment of a given case.

1. Refraction.
2. Occlusion.
3. Orthoptic Training.
4. Operation.

REFRACTION. This is necessary in order that any refractive error may be corrected and has to be carried out under the full action of atropine. Atropine sulphate 1 per cent (drops or ointment) is ordered to be used to each eye twice daily for three days prior to the examination. Parents should be warned that children sometimes show the effects of undue systemic absorption of atropine (restlessness, flushing, dry skin and delirium) and that the administration should be stopped if there is any suggestion of this. Some newer synthetic mydriatics (e.g. Cyclopentolate, Mydrilate), which have a quicker and less prolonged action, are now tending to replace atropine, the effects of which may take ten days to disappear.

While the pupils are fully dilated the eyes are examined to exclude the presence of disease, and glasses are ordered, if necessary.

If the squint is a recent one, and the error of refraction appreciable, spectacles may control the squint completely and no other treatment may be required. The glasses must be worn constantly from morning to night, whether the child be at home or at school, and must be repaired or replaced without delay if they are damaged. As the glasses control the excess accommodation, the eyes are now in a position for the development of binocular vision to follow its normal course and they may be able to be discarded in later years, when full binocular function has been attained. Figure 37 shows a case of accommodative convergent squint with and without glasses.

OCCLUSION. Even with spectacles which adequately correct the refractive error, there is no chance of binocular function developing while the vision of the squinting eye is less good than that of its fellow. Occlusion of the 'straight' eye is, therefore, employed to force the squinting eye into activity.

Neither the child nor his parents take kindly to occlusion at first. The child because the vision in the squinting eye is relatively poor, and the parents because they usually do not understand why the child should be forced to use a less effective eye and thus have to face schooling and the difficulties of everyday life with vision of a less high grade than that to which he has previously been accustomed. It is for these reasons that patient explanation must be made to the parents as to the reasons behind the occlusion. They must be shown that the occlusion is to be worn constantly and are assured that the amblyopia will soon begin to improve. Some warning as to extra care in handling the child in such situations as in crossing the road would not be out of place, and a message

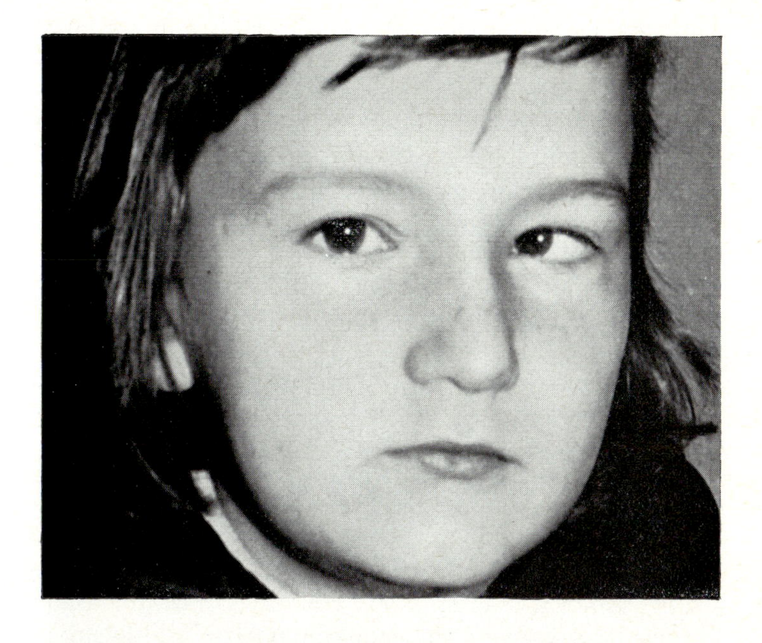

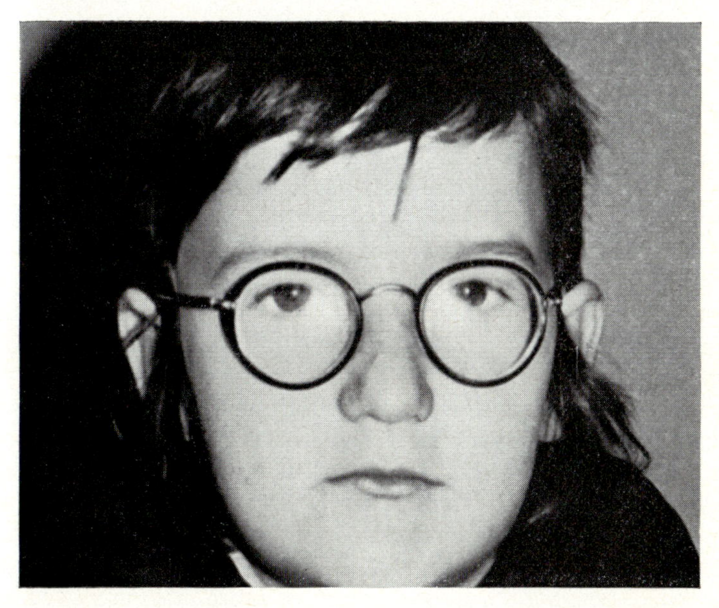

Fig. 37 Accommodative squint, with and without glasses.

must be sent to the teachers of school-age children to explain why the standard of school work may fall temporarily. The child himself needs constant encouragement.

In the early stages occlusion is best achieved by the use of a circle of zinc oxide plaster over the eye, with a strip of greased lint or gauze to prevent the plaster sticking to the eyelashes. As the child becomes accustomed to occlusion, the plaster is replaced by a 'blinker' of opaque material fastened to the glasses.

The duration of occlusion depends on the results achieved and on the degree of original visual depression. If improvement is not seen within six or eight weeks, and one can be sure that effective occlusion is being carried out, the likelihood is that the amblyopia is incurable. In a successful case this treatment is persisted in until the vision in the two eyes is equal, when it is not uncommon to find that the squint becomes 'alternating', and the child uses either eye with equal facility. With the development of alternation, it will be necessary to explain to the parents why the squint has switched into the originally straight eye and why the angle of squint sometimes increases when the amblyopic eye begins to take up fixation. It must be stressed that treatment of the amblyopia is the primary object and that it is designed for the preservation of vision.

Some squints will be cured by the provision of glasses together with occlusion to remove the amblyopia.

ORTHOPTIC TRAINING. Although not a new technique, it is only during the past twenty years that the science of orthoptics has taken a full place in the system of treatment of squints. Its practitioners are now recognised medical auxiliaries with their own requirements for training, examination and registration. Orthoptics is best looked upon as 'ocular physiotherapy' and is concerned with restoration of binocular function, or assisting its development, if this has been delayed or prevented by the presence of a squint. While not the 'be all and end all' of squint treatment, orthoptic examination assists the surgeon in deciding in which cases operation is indicated and at what stage of the treatment, and shows what degree of binocular vision may be expected to be achieved in a given case. Certain squints, particularly those of small degree and of recent onset, may be cured by a properly designed course of exercises. The surgical results in those cases for which operation is indicated may be appreciably improved by pre-operative and post-operative orthoptic training.

OPERATION. As orthoptic training is physiotherapy as applied to the visual functions, so squint surgery is 'orthopaedic surgery as applied to the eye'. Save in so far as its purpose is to place the eyes in the most effective position for the development of correct visual acuity, it is no cure for amblyopia.

Surgery is reserved for those cases of squint in childhood which are not cured by the provision of correct glasses, together with occlusion and orthoptic training; those children who are too young to co-operate in training; and those cases in which reduction of a very large angle of squint is required in order that more effective orthoptic training may be given. It is also the only possible line of treatment in neglected squint, both in adults and in those children showing incurable degrees of amblyopia, when the problem is a purely cosmetic one.

Although a description of surgical technique is not proposed, it can be said that the principle of operation is to achieve parallelism of the visual axes by planned lengthening or shortening of the ocular muscles. The operation is carried out under general anaesthesia and, as a rule, the duration of stay in hospital is of the order of a day or two, and it is common for no pads or bandages to be used post-operatively. No child is too young for squint surgery if the nature of the condition demands it, and it is now thought that operation at an early stage is often the best treatment in those cases where the squint is of early onset. To place the eyes in a parallel position will in some cases allow the development of normal binocular reflexes.

It is often necessary to carry out the operative treatment of squint in stages; and parents have to be warned that the deformity may not always be rectified by a single wave of the surgical wand. If this is understood, a certain amount of disappointment may be avoided.

To summarise the modern treatment of squint, it can be said that patients can always be promised a pair of straight eyes. Whether they will simply be cosmetically straight or will also work together as a team in the function that is 'binocular vision' depends on a number of variable factors, not the least important of which are the early application of adequate treatment and the perseverance of both parents and child in what is often a tedious and sometimes a disappointing process.

DIVERGENT SQUINT. Although convergent squints are about ten

times as frequent as divergent squints, there are one or two features of the latter which are of importance.

Divergent squint is classically an intermittent deviation and generally shows a later age of onset than convergent squint. It is only seen on distant gaze and is much more obvious under conditions of bright illumination, the child often closing one eye in bright light, to avoid the diplopia resulting from the squint.

These squints respond poorly to orthoptic treatment and the majority require surgery. The prognosis for treatment in divergent squint is very good and these children only rarely become amblyopic.

PARALYTIC SQUINT

The causes of acquired squint are many, and include head injuries, intracranial vascular accidents, aneurysms and tumours.

Symptoms

The only symptom, and a very distressing one, is *diplopia*, which may lead to giddiness and nausea. It is usually worst in one particular direction of the gaze, and the patient will sometimes be able to overcome it by the adoption of a head posture which avoids the necessity of looking in that direction. If relief is not possible in this way, it will be obtained by closing or covering one eye.

Diagnosis

A frank squint, or definite limitation of movement of an eye, may be obvious, in which case the diagnosis will probably be apparent. In lesser degrees of weakness, or incomplete paralysis, more refined methods of investigation are required.

Applied Anatomy

The three motor nerves to the eye muscles are distributed as follows:

III	Levator Palpebrae Superioris
	Superior Rectus
	Medial Rectus
	Inferior Oblique
	Inferior Rectus
IV	Superior Oblique
VI	Lateral Rectus

If, therefore, one can determine which muscle is at fault, the nerve involved can be deduced.

Which Muscle is Involved?

To answer this question, two further questions are considered.

1. In which direction of the gaze does the maximum separation of the images occur?
2. To which eye does the 'farthest away' image belong?

The reasons behind these questions are that the maximum separation of images occurs in the direction of action of the weak muscle, and that the image belonging to the affected eye appears to be farther away from the patient than that belonging to the normal eye.

The Six Diagnostic Positions

For certain anatomical reasons, the details of which are outside the scope of this work, the activity of the various extraocular muscles can be separately tested by asking the patient to look in six directions. These are set out below. Using a torch as a fixation object, it is possible to find the position in which the two images are farthest apart. From the table which follows, it will be seen that maximum diplopia occurring in any of these positions points to the involvement of one of a pair of muscles, one belonging to each eye. Having found the position of maximum separation, the eyes are covered in turn and it is possible to find out to which eye the farthest displaced image belongs. This is the eye with the weak muscle.

The Diagnostic Positions

1. Horizontally to the right	Right Lateral Rectus	Left Medial Rectus
2. Horizontally to the left	Right Medial Rectus	Left Lateral Rectus
3. Up and to the right	Right Superior Rectus	Left Inferior Oblique
4. Up and to the left	Right Inferior Oblique	Left Superior Rectus
5. Down and to the right	Right Inferior Rectus	Left Superior Oblique
6. Down and to the left	Right Superior Oblique	Left Inferior Rectus

This table can be expressed diagramatically and is perhaps more readily referred to in this way. The circles represent the corneae and the scheme is drawn from the viewpoint of the observer facing the patient.

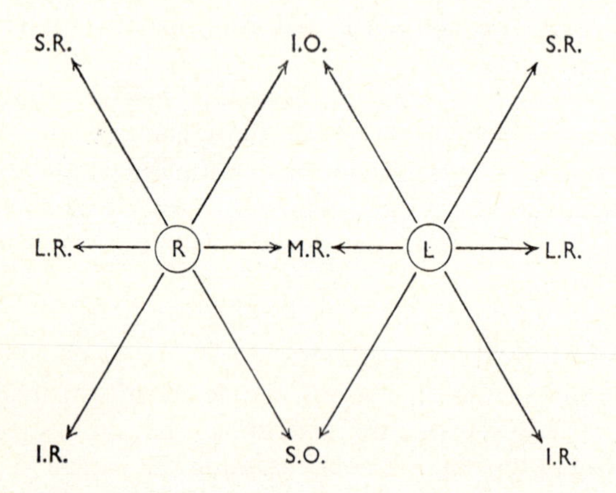

By the use of such a scheme it is usually possible to answer the two diagnostic questions.

1. In which direction of the gaze does maximum separation of images occur?

2. To which eye does the 'farthest away' image belong (Fig. 38 and 39)?

The Management of Paralytic Squint

In the early stages, and to avoid the distress of diplopia, the only possible treatment is occlusion of one eye. The eyes should be occluded alternately, in order to prevent the development of ocular neglect. In some cases, and particularly when the palsy is incomplete, diplopia can be avoided by turning the head into such a position that the paralysed muscle is not called upon to act.

The timing of the institution of more active treatment depends on the cause of the condition and on the degree of natural recovery which is taking place. In some cases, especially those due to trauma, gradual improvement may be seen over a period of eight to twelve months. During this period orthoptic exercises may help in the stimulation of the weak muscle and the prevention of secondary contractures.

When a static position has been reached, it may be found that the patient is avoiding the irritation of diplopia by the adoption of a slightly abnormal head posture. Alternatively, the angle of deviation may be sufficiently small to warrant the provision of spectacles incorporating prisms. These will often permit the

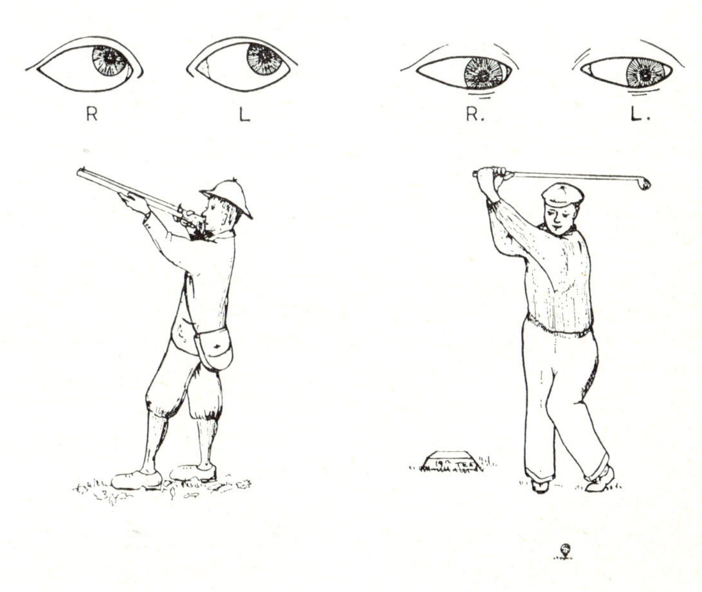

Fig. 38 Ocular movement up and to the left=R. Inf. Oblique and L. Sup. Rectus.

FIG. 39 Ocular movement down and to the left=R. Sup. Oblique and L. Inf. Rectus.

recovery of binocular vision, at least in some positions of the gaze. In a final group of cases, surgical adjustment of the position has to be attempted and, even though it may have to be carried out in stages, can usually be relied on to produce binocular vision, at least in some directions of the gaze. Efforts are directed at obtaining single vision in the 'straight ahead' position and in the lower field of vision. Diplopia above the horizontal plane is much less of a hardship than below, when the difficulties involve problems with reading, coping with steps and stairs, and so on.

LATENT SQUINT (HETEROPHORIA)

As has been said, the factor which ultimately controls the position of the eyes is the stimulus of fixation; that is, the desire to bring the image of the object regarded on to the macula of each eye, as this is the retinal area at which most acute vision is possible. But there is also, once the normal binocular reflexes have been developed, a mechanism acting on the ocular muscles themselves, which tends to maintain the direction of the gaze, even if the fixation reflex is not active. Such a state of affairs is present if one eye is occluded while fixation is maintained with the other. The covered eye takes up what may be called 'the position of rest' and resumes fixation when the cover is removed. Under ideal conditions the covered eye remains directed as before and makes no movement of recovery when it is once more uncovered (orthophoria).

In most individuals this ideal is not attained and the covered eye takes up a position of slight divergence (exophoria) under cover, moving to take up fixation once more when the cover is removed. In most instances, the slight muscular effort required to maintain binocular vision is not sufficient to produce symptoms.

If the degree of imbalance is considerable, the neuro-muscular effort required to maintain binocular fixation may be sufficient to provoke headache or various symptoms of ocular discomfort. Occasionally, under conditions of ill-health, a latent squint may break down and become a frank strabismus, and thus account for the appearance of a squint in an adult, with intermittent diplopia as a presenting symptom.

Ocular symptoms caused by the presence of a latent squint can often be relieved by orthoptic treatment, the object of which is, in this case, to strengthen the control of the ocular movements.

Convergence Insufficiency

While not strictly a state of ocular imbalance, this quite common condition deserves consideration here. There is an inability to sustain convergence and this is manifest clinically as blurring of print while reading, 'running together' of words, and a sensation of eyestrain. The patients are usually young or middle-aged, often admit to not being very fit, and are sometimes convalescent after an illness. The condition is also seen as a manifestation of anxiety states.

Convergence insufficiency is one condition in which orthoptic treatment is an almost certain means of relieving symptoms. A short course of training is combined with exercises to be used at home. Improvement of convergence takes place in a short time.

CHAPTER ELEVEN

Glaucoma

The normal intraocular pressure varies, but on the average it is approximately 17 mm of mercury. When this pressure is disturbed and the eyeball becomes abnormally hard, damage to the nervous and vascular tissues at the posterior pole of the eye occurs, and visual defects result therefrom.

Under normal circumstances the intraocular pressure is the result of the balance between the inflow of fluid into the eye and its outflow through the drainage channels. The site of the formation of aqueous humour is the ciliary body, whose function is apparently comparable with that of the choroid plexuses in the ventricles of the brain. From the ciliary body the fluid passes into the anterior chamber of the eye from which it is absorbed and gains access into the blood stream. There is no absolute certainty as to the route taken by the aqueous humour from the anterior chamber. Part passes into the canal of Schlemm which lies in the junction between cornea and sclera, and from here it drains into the episcleral and probably conjunctival vessels. Whether or not this is the only drainage channel is not clear, for it has been suggested that the iris and the choroidal blood vessels also provide a means of outflow for the aqueous. The exact part which is played by the various factors in the production of a glaucomatous state is still largely unknown, but obstruction at the angle of the anterior chamber is now considered to be of prime importance in most cases.

Classification

1. Infantile.
2. Adult, which is subdivided into
   (*a*) Simple (open-angle glaucoma)
   (*b*) Congestive (narrow, or closed angle glaucoma).
3. Secondary.

For Anatomy of the Eye see Figure 1.

INFANTILE GLAUCOMA (BUPHTHALMOS)

Due to congenital abnormalities of the drainage channels, the intraocular pressure in the infantile eye may be abnormally high. The young eye responds in a different way from that of an adult to this rise of pressure, for the coats of the eye in an infant are not rigid and become distended by the abnormal pressure within them. The result is that the eye becomes larger than normal. It is usual for attacks of increase of pressure to occur; these are usually marked by pain as manifest by increased irritability of the child and a tendency to rub the eye, by redness of the eye and by a steamy appearance of the cornea, together with lacrimation. In between attacks the eye may appear normal apart from a tendency to be somewhat large in size. The condition may by unilateral or bilateral. If untreated, the tissues at the posterior pole of the eye, particularly the optic nerve head, become atrophic, and vision is lost.

Treatment. Medical treatment with miotic drops is seldom successful in controlling the condition and surgery is required. This surgical treatment may consist of an attempt to re-open the drainage channels and the angle of the anterior chamber (goniotomy or cyclodialysis) or to make a new filtering channel from the anterior chamber (trephine, iridencleisis). The treatment of buphthalmos is difficult, presumably because part at least of the damage to the eye has been done before the patient is brought for advice and perhaps even during the intrauterine period.

ADULT GLAUCOMA

Simple Glaucoma (Open-angle Glaucoma)

In this condition the elevation of intraocular pressure is never sufficiently great to cause episodes of severe visual loss as in the case of congestive glaucoma. The major difficulty with regard to diagnosis, therefore, is that a considerable amount of damage may occur without there being any major symptoms to bring the condition to the patient's notice. A large proportion of sufferers from simple glaucoma who present for examination with a request for a routine eye test have no suspicion of any abnormality of the eyes; though it is characteristic of glaucoma for abnormally frequent changes of spectacles to be required owing to rapid changes of refraction. The condition is characteristically bilateral, and it may well be that a large part of the visual field of one eye is lost and the second eye seriously affected before the patient requests advice. It is the interference with the visual field which is the primary visual change in simple glaucoma. Central vision, that is, visual acuity as normally recorded, is often affected very late, and it is this preservation of central vision which makes for the fact that the disease may be considerably advanced before it is detected.

Although the fundamental aetiology of simple glaucoma is obscure, there is no doubt that the primary result of the raising of the intraocular tension is an atrophy of the nerve fibres at the optic nerve head. This leads to gradual changes in the visual field affecting both the peripheral limits of the field of vision and also certain bundles of fibres in the central region. Clinically, these effects show themselves by the presence of visual field changes, the earliest of which requires special detailed measures for its detection, and the presence of a particular type of atrophy of the optic nerve head. This atrophy, which leads to 'pathological cupping' of the optic nerve head, is typical of glaucoma, and it comprises an excavation of the nerve head with resultant depression of the blood vessels as they pass over the edge of the optic disc together with abnormal visibility of the foramina in the lamina cribrosa through which the bundles of nerve fibres pass. The cupped optic disc is also abnormally pale due to the presence of optic atrophy (Fig. 40). The difficult distinction which has to be made is between this type of optic nerve appearance and the physiological cupping which occurs in the normal nerve. This physiological cup does not extend to the edges of the optic disc and is not associated with optic atrophy (Fig. 6). However, in some cases it is on examination of the visual field that a diagnosis of glaucoma simplex will rest, together with examination of the intraocular pressure. This last examination is made with an instrument called a tonometer which gives an approximation of the degree of pressure within the eyeball. As there is an appreciable diurnal variation in this pressure, the test may have to be repeated at intervals throughout the day in order that the maximum reading may be obtained. Even repeated tonometry may not give an abnormal reading in every case, and there is no doubt that some eyes suffer damage characteristic of glaucoma while the intraocular pressure is never above normal levels. In these cases there is presumed to be an additional factor involved.

The diagnosis of glaucoma simplex, therefore, may depend upon specialised techniques, and there may be no definite symptom

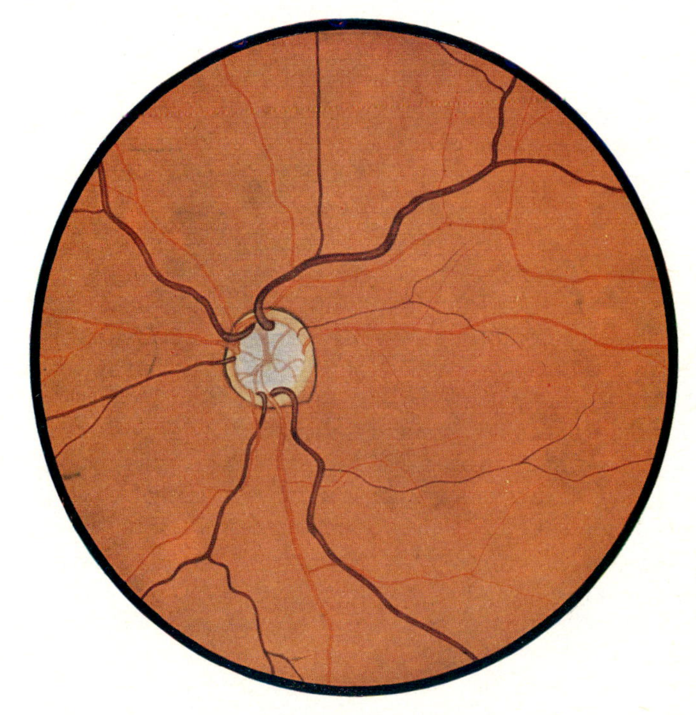

Fig. 40 Glaucomatous cupping of the optic nerve head.

which suggests the presence of disease either to the patient or to the practitioner. It is this absence of specific symptoms which has led to the present position of glaucoma simplex being one of the chief blinding diseases of the elderly, the early detection and treatment of which is of great concern to the ophthalmologist. It is now generally accepted that this disease is present in between one and two per cent of the population over the age of forty, and that this proportion rises with increasing age. There is also no doubt that simple glaucoma often has a familial tendency and the relatives of known sufferers from glaucoma should be advised, at least after middle age, to undergo an annual examination of the eyes. In a disease of insidious character such as this, responsibility for early detection lies in the first place with the refractionist who will often be first consulted with regard to change of spectacles. Although tonometric studies of intraocular pressure cannot at present be included in every routine eye examination in adults, refractionists must be suspicious of any sudden change in a patient's eyes, must be aware of the tendency for glaucoma to run in families, and must refer for detailed examination any case presenting suspicious features, particularly abnormalities of the optic disc. Any visual field loss which has already occurred before treatment is instituted cannot be recovered as the optic atrophy is permanent and irremediable.

There is a further point which should make us suspect chronic glaucoma, and this is the occurrence of retinal vein thrombosis. It is now recognised that a significant number of cases of glaucoma have venous thrombosis as their presenting feature. It should once more be pointed out that the long-term use of steroids locally may induce glaucoma in some eyes.

Treatment. In the treatment of glaucoma simplex the miotics (eserine and pilocarpine) represent the first line of attack. The use of these drugs maintains a small pupil and assists the outflow of fluid from the eye. In a proportion of cases the intraocular pressure can be kept at normal levels and further visual field loss prevented. The strength and frequency of application of the drugs depends on their effect in the individual case. In most cases pilocarpine, 1 or 2 per cent, used three or four times a day will maintain control of the condition. Other drugs which may be used are phospholine iodide which is a long-acting miotic, and adrenaline (Eppy, 1 per cent) which is often effective in combination with pilocarpine. The systemic carbonic anhydrase inhibitors (Diamox, acetazolamide) are not as useful for long-term use in chronic glaucoma as they are in congestive glaucoma.

The next important step is to ensure supervision of the case. Repeated examination of the state of the visual field and of the intraocular tension is essential, for it is only on these that an assessment of the adequacy of control can be based. In particular, in unilateral cases, a careful watch can be kept on the opposite eye, in order to detect the earliest signs of disease here.

If miotics fail to control the disease a drainage operation must be carried out, in order to increase the outflow of aqueous humour from the eye.

Congestive Glaucoma (Narrow-angle Glaucoma)

ACUTE CONGESTIVE GLAUCOMA. This is one of the emergencies of ophthalmology. The patient, who is usually elderly, complains of a sudden onset of pain and blurring of vision in the eye. This pain may be so acute as to produce collapse and vomiting and has been known to mislead the medical attendant into believing that some general or abdominal cause of vomiting was present. The pain is located mainly in the eye, but spreads into the distribution of the fifth cranial nerve, giving rise to referred pain in the brow,

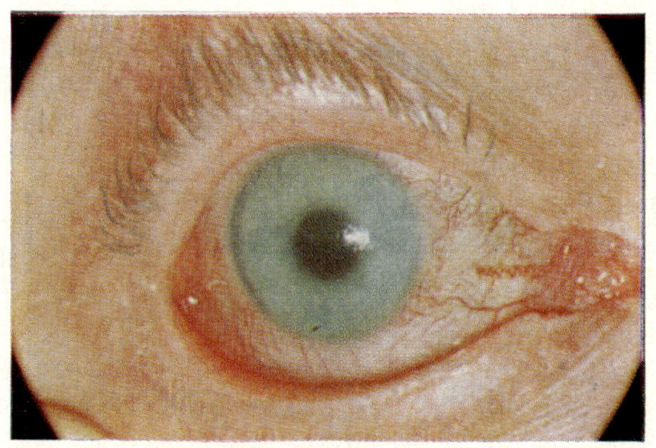

Fig. 41 Acute glaucoma.

the temple and the cheek. Visual acuity is much reduced and examination reveals that the eye is congested, the cornea hazy due to interference with its nutrition, and the pupil semi-dilated and inactive (Fig. 41). The anterior chamber is usually reduced in depth. The eyeball is tender to the touch, but it should be possible to palpate the globe through the closed lids, when it will be found that the intraocular tension is raised and the eye feels stony hard. Additional diagnostic assistance can sometimes be derived from the history, for the typical attack of congestive glaucoma comes on in the evening, precipitated perhaps by some emotional upset or over-exertion.

In most cases there is a history of previous visual disturbance. These attacks consist of blurring of vision and some ocular discomfort, worse in the evening and relieved by sleep. The attacks are often regular in their occurrence and may be associated with the appearance of coloured 'haloes' around lights at night.

The attack of acute congestive glaucoma may resolve spontaneously but, if it persists, it will progress to serious interference with the vision.

Treatment. Treatment comprises intensive use of miotic drugs in an effort to constrict the pupil and to increase the outflow of aqueous humour. Eserine can be used in a strength of 1 per cent every fifteen minutes during the first hour or so, together with hot bathings of the eye in order to increase the effectiveness of the drug. The patient should be treated in bed and general sedation is valuable to ensure rest and it seems to assist the reduction of intraocular tension. The use of Diamox has now become established in the treatment of acute glaucoma. An initial dose of 500 mg is followed by 250 mg six-hourly. If the condition is going to respond to the miotic it will usually do so during the first few hours of treatment, when the cornea will clear, the pupil will contract and the vision improve.

A further measure which will often help to control an acute attack of glaucoma is the use of systemic osmotic agents to draw fluid out of the eye. The most useful of these agents is glycerol, given as a fruit-flavoured drink in a dose of 1·5 g/kg body weight. If medical treatment fails to produce improvement in the congestive condition (and the patient is best observed in hospital during the treatment, if this is possible), then surgery has to be considered. This surgical treatment usually consists of an operation upon the root of the iris, either a simple iridectomy or an iridectomy combined with an operation to produce a new filtering channel for the absorption of the aqueous humour. This operation will nearly always prevent recurrence of the condition.

If medical treatment is successful in controlling the acute attack, a prophylactic iridectomy is usually advised in order to prevent further attacks. Owing to the frequency with which the condition becomes bilateral, most surgeons would advise operation on the second eye also.

SUB-ACUTE OR CHRONIC CONGESTIVE GLAUCOMA. In chronic congestive glaucoma there are no fulminating attacks in which major symptoms are present, but there are constantly recurring attacks in which the intraocular pressure is raised and in which the vision is to some extent interfered with. These attacks occur characteristically at night, or after a visit to the cinima, when the patient will complain of blurring of vision, and will see coloured 'haloes' around lights. These haloes are due to corneal oedema and are evidence of abnormally high intraocular pressure. Between the attacks the eye may appear quite normal, with full vision and a normal intraocular pressure as tested with the tonometer. The attacks are more common in the winter than in the summer, and when they occur they usually occur each night, often being associated with headache or a sense of fullness in the eye. If the intraocular tension is recorded during an attack of haloes it is found to be raised.

Although the vision returns to normal between the early attacks of congestive glaucoma these continual recurrences cause gradual damage to the eye. The iris becomes adherent to the posterior surface of the cornea at the periphery, and the constantly recurring attacks of elevated intraocular tension lead to cupping of the optic nerve head and to consequent visual field loss. Although chronic angle-closure glaucoma has been described as an entity separate from acute glaucoma, it must be made clear that the two conditions are fundamentally the same. Although acute glaucoma may be the first evidence of disease, it often occurs in a patient who for months or years has suffered from recurrent episodes of 'haloes' and blurring of vision, which represent attacks of closure of the angle of the anterior chamber each of which has resolved before the intraocular circulation has become obstructed with the production of an acute attack.

Treatment. Although the attacks can be controlled with miotics (eserine and pilocarpine) most now consider that a small iridectomy is the best insurance against further attacks, and this is an operation almost without risk. It is only in those cases where the patient is unwilling for surgery and can remain closely under supervision that continuation of medical treatment is advised.

The importance of complete control of the ocular tension, and the insidious nature of the disease process, have led to the establishment of glaucoma clinics which the patients attend at regular intervals for examination. The earliest signs of deterioration can then be detected and a strict system of follow-up supervision maintained.

Secondary Glaucoma

The intraocular pressure sometimes becomes raised in the course of other diseases of the eye, and this possibility has to be considered when the patient complains of an increase in pain or of visual defect.

The possible causes of this condition are many and include injuries (particularly if associated with intraocular bleeding), iridocyclitis, intraocular tumours, and the late effects of a retinal vein thrombosis.

The treatment is that of the underlying condition.

CHAPTER TWELVE

Injuries

Certain injuries, such as those due to chemical burns, have been discussed in relation to the parts of the eye affected, but it is convenient to consider the effects of contusions and penetrations of the eye as a whole.

CONTUSIONS

Without Penetration

Such injuries are caused by blunt objects, the commonest of which are fists, blows during sport and flying objects at work.

The effects of such injuries upon the eye vary from a simple corneal abrasion to a gross disorganisation of the globe.

The cornea may be abraded or it may simply show diffuse haziness following oedema due to the injury. Such oedema resolves without special treatment during the course of a day or two.

There may be sub-conjunctival haemorrhage from rupture of a small conjunctival blood vessel but, except in so far as it may conceal an injury to the deeper parts of the eye, such haemorrhage requires no specific treatment.

In the anterior chamber there may be bleeding. If slight, this shows itself immediately after the accident as a diffuse haziness of the aqueous humour with resultant difficulty in visualising the details of the iris or the fundus. In a few hours the blood settles to the lowest part of the anterior chamber where it lies and shows the characteristic fluid level. The condition is known as *hyphaema* (Fig. 42). Sometimes the anterior chamber becomes literally filled with blood and this may lead to secondary glaucoma.

There is some discussion as to whether or not cases of traumatic hyphaema should be treated with a mydriatic, for it is said that there is some risk of precipitating fresh bleeding from the torn iris vessels by the use of atropine, but as there is always some degree

For Anatomy of the Eye see Figure 1.

of traumatic iritis present in these cases, it is considered that the use of atropine 1 per cent twice daily is justifiable and desirable. The patient should be kept at rest until the haemorrhage has been absorbed.

The iris may show other effects of the injury. There may be *traumatic mydriasis*, a dilatation of the pupil following interference with the nerve supply of the sphincter from distortion of the eyeball at the moment of impact. The mydriasis usually passes off, but may be permanent, presumably due either to gross interference

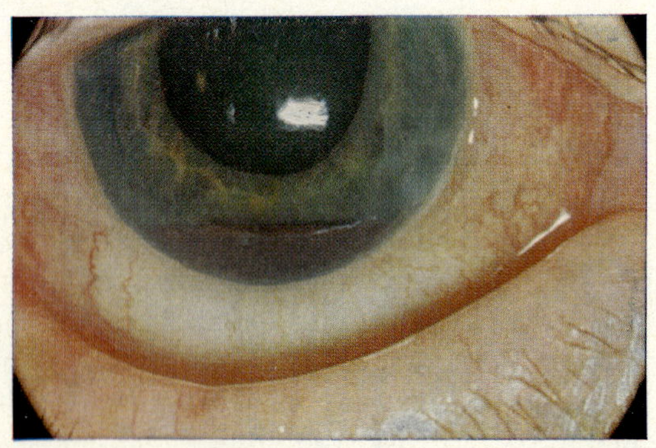

Fig. 42 Hyphaema.

with the ciliary nerves or to tears at the pupil margin, these tears interrupting the continuity of the sphincter. Other tears may arise at the periphery of the iris, the iris being torn away from its root in the condition known as *iridodialysis*. If the dialysis is small and in the upper part of the iris where the defect is covered by the upper lid, the patient may complain of little disability. If the defect is large and in the lower part, then surgical repair may have to be attempted owing to the visual defect which results from the deformity.

Various types of cataract follow non-penetrating contusions of the eyeball. They are due to interference with the nutrition of the lens from damage to the ciliary body. These cataracts may occur in the period immediately following the injury, but they are sometimes delayed and may be the cause of subsequent visual defect. In more gross injuries the lens of the eye may actually be displaced from its attachments to the ciliary body, either as a *partial dislocation*, or as a *total dislocation*, when it may pass forwards through the pupil to the anterior chamber or backwards into the vitreous. Suspicion of serious injury to the eye will be raised in such cases by the presence of gross visual defect.

In the posterior part of the eye, there may be bleeding from the retinal vessels into the vitreous or disturbance of the retina itself due to contusion. The retina suffers from the momentary distortion of the eye which takes place at the time of impact and its nutrition is interfered with so that oedema occurs, which, if in the central area, may interfere considerably with vision. The condition is known as *commotio retinae* and is seen ophthalmoscopically as a greyish pale area of the fundus in which retinal detail is obscured and where there may be some small retinal haemorrhages. The natural tendency of the condition is to recovery and this is assisted by the use of atropine which ensures ocular rest, and by keeping the patient at rest for a few days. Recovery is usually complete, but there are cases of severe retinal contusion in which permanent pigmentary disturbance results and in which visual acuity may be seriously interfered with. Another retinal disturbance which may sometimes follow such blunt injuries is *detachment of the retina*, in which a retinal tear develops at the time of the injury and allows fluid to obtain access to the under surface of the retina leading to detachment of the retina from its bed. The patient will complain of increasing failure of vision, often delayed. Treatment is surgical.

The choroid also may be torn as a result of distortion of the eye; the tears show themselves as crescentic white scars, concentric with the optic disc and usually situated towards the posterior pole of the eye. They represent areas of choroidal atrophy through which the sclera shines white.

With Rupture of the Globe

These more serious injuries follow a contusion of such a severity that the coats of the eye are unable to withstand the pressure developed within them. The commonest place for rupture to occur is at the junction of the cornea and the sclera. There will usually be haemorrhage within the eyeball, careful examination in a good light will reveal the rupture, and there will probably be intraocular contents presenting in the wound. Vision is grossly interfered with and these injuries are not usually to be mistaken.

The only warning that should be raised is that a severe penetrating injury to the globe may be overlooked if there is also injury to the lids or fracture of orbital bone. If the lids are oedematous and swollen or perhaps torn, active steps must be taken to inspect the eyeball and if possible to assess the visual acuity. It is only in this way that severe injury to the globe can be excluded.

In any ocular injury, whether due to contusion or to penetration, some assessment of visual acuity should be made at an early stage. Not only will this enable one to suspect the presence of severe injury to the eyeball, but it may also be important from the medico-legal point of view, when the question of visual damage may be raised.

PENETRATING INJURIES

Under this heading are included penetrations of the eye due to pointed objects, such as scissors, knife blades, bows and arrows and so on, and also those which are caused by flying particles, particularly in industrial processes, and, nowadays, in motor accidents.

The possible causative instruments are many, and the injuries resulting from them are divided most readily into those in which no foreign body is retained within the eye and those in which there is a foreign body retained.

In the first group there are injuries due to sewing needles, points of knives, sharpened sticks, and bows and arrows. The resulting injury is usually of a major degree, though lesions occurring as a result of penetration with such things as fine wire are occasionally very small and difficult to see. In most cases the patient complains of pain and gives a history of something having struck the eye. The iris is often deformed if the wound is in the anterior part of the globe and it may be actually prolapsed on to the surface through the penetration. The wound may be anywhere in the cornea or corneo-sclera and the edges of it will often be visible (Fig. 43). Examination of the deeper parts of the eye is often impossible on account of photophobia and interference with the clarity of the intraocular structures, particularly the lens, which, if it is involved in the accident, rapidly becomes opaque and prevents an adequate view of the interior of the eye. The main difficulty in this type of injury is the fact that a large number of them occur in children in whom examination of the eye is notoriously difficult and in whom the spasm of the lids is often sufficient to prevent detailed examination without special equipment. It should also be remembered that the child may, through shame of the incident which led to the injury, conceal the severity of the damage or even the fact of having received an injury. It may not be until the eye becomes red and painful during the subsequent few days that the true nature of the condition is appreciated.

Treatment of these wounds is a matter for the surgeon, and it involves repair of the wound if this is possible, toilet of the extruded

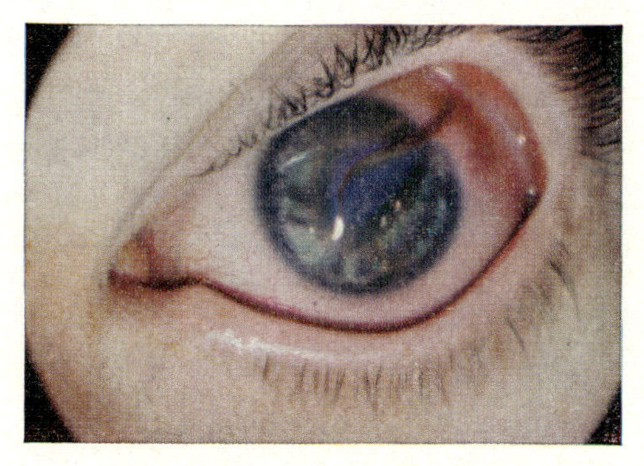

Fig. 43 Laceration of the cornea.

intraocular contents and protection of the damaged tissues by a plastic flap provided by conjunctiva. If the lens remains intact, and if the wound settles down and heals without infection, the visual prognosis is quite good, particularly in those cases in which the central part of the cornea has not been extensively damaged by scarring. But a number of eyes each year are lost or damaged as a result of injuries by such childhood games as bows and arrows, cowboys and Indians, dagger fights and so on. Parents, on the whole, do not seem to be aware of the risk involved in this kind of play. It is natural for small boys to shoot arrows at one another and they naturally also aim for the head, but those who have control of children should, I think, be reminded constantly by those of us who have the responsibility of advising them that these games are potentially dangerous, and the common result of an ocular injury is for the eye either to be lost, or to be rendered at the best partially-sighted.

It seems convenient at this point to mention the question of fireworks. Fireworks of all sorts are sold over the counter with very little apparent control and there are, each year, a number of serious ocular injuries due to explosions from these. The common story is that the youngster picks up the firework which seems not to be fused properly and he often blows on it in order to stimulate its activity. The resultant explosion not infrequently destroys an eye and occasionally destroys two. The responsibility of adults in the case of firework celebrations does not end after the explosives have been bought from the shop and passed over to the youngster. They should be constantly supervised in use and the potential danger appreciated.

Penetrating Injuries with Retention of Foreign Body

These injuries occur almost always in industrial processes or during mechanical operations. High speed drilling and grinding machines, pneumatic riveters and a number of engineering processes are very often at fault and, although workers are provided with goggles they complain that the range and acuity of their vision is interfered with by the use of goggles and they often fail to protect themselves adequately. Probably the commonest source of this type of penetrating injury, however, is the ordinary hammer and chisel. Chisels are allowed to become 'mushroomed' at the head and the edge of the hammer is scarred. Unless the chisel is kept in good condition by regular attention at the grinding wheel, a glancing blow at its top will send flakes of steel at high speed through the air and such flakes may penetrate the eyeball.

These flying particles are often extremely small, and travel through the air at high speed. As a result they are usually sterile. They enter the eye through a small wound and the workman may sometimes be scarcely aware of the incident. He may feel simply that he has 'something in his eye'; and this makes diagnosis difficult and makes it necessary that medical officers dealing with men engaged in this kind of process should be constantly on their guard and on the lookout for an injury of this kind, For if a piece of metal remains within the eye, although not at first producing serious visual defect it will in the subsequent months or years induce chronic chemical irritation and will ultimately destroy the sight. This condition is known as *siderosis bulbi* and results from the retention of a ferrous particle within the globe.

Examination of the injured eye may show a wound of entry, perhaps with prolapse of iris, and such an injury is not likely to give rise to difficulty in diagnosis. It is the injuries in which the wound of entry is small and perhaps lying at the margin of the cornea, or even in the sclera where it is covered by conjunctiva, that the true nature of the condition may be missed. Careful examination in good light will often show a corneal opacity and it may be possible to see a penetrating wound in the iris. A hole in the iris following a history such as has been outlined is practically diagnostic of the retention of a foreign body within the eye. Visual acuity may be good at first, but it will become gradually less so as time goes on if the lens of the eye has been damaged, for traumatic cataract usually develops during the subsequent two or three days. Before development of such cataract, however, it may be possible to visualise the foreign body within the globe or to see the evidence of intraocular disturbance such as vitreous opacities or intraocular haemorrhage.

The only sure way of arriving at a diagnosis is by X-ray, and it is a matter of routine for any patient who gives a history suggestive of penetrating injury of the eye to have his orbits X-rayed to exclude a radio-opaque foreign body.

The treatment of these cases is similar to that of those showing a penetrating injury without retention of a foreign body, in that the wound of the eye has to be repaired and detailed X-ray studies are taken in an effort to localise the site of the foreign body. If it is considered to be magnetic, it can then be removed either through the wound of entry or through a separate opening made at the back of the eye.

Sympathetic Ophthalmia

Although sympathetic ophthalmia is not a common condition, it occasionally occurs as a result of a penetrating injury to an eye, when the injured eye fails to settle down in the course of the subsequent two or three weeks. There is a resultant irritation of the second eye in these cases and this, the sympathising eye, develops a chronic low-grade iridocyclitis which may be extremely destructive, and the result of this bilateral uveitis may be the loss of both eyes. The condition occurs only as a result of a penetrating injury to the globe and does not develop during the first two or three weeks after the accident, so it is safe for these eyes to be watched during that period. If it is considered that the injured eye is not likely to be valuable from the point of view of vision and that the inflammation is failing to respond to treatment, then excision of the affected eye has to be considered in order to reduce the risk of involvement of the second eye. The condition of sympathetic ophthalmia is thought to be due to a sensitivity reaction developing in the uveal tract of the uninjured eye, and it can to some extent be brought under control by the use of steroids. Despite the availability of steroids, however, the decision as to whether or not an injured eye should be preserved is always one giving rise to a degree of anxiety and must be taken by a specialist ophthalmologist.

Injuries to the Orbit

Orbital haematoma may follow contusion injury to the orbit, and fractures of the orbital bones may lead to displacement or damage of the extraocular muscles. Interference with the ocular muscles gives rise to diplopia, sometimes very incapacitating. No treatment is advised in the early stages of an injury of this kind, save realignment of any misplaced bone, if this can be achieved. A period of some months often passes before the oculomotor position becomes stabilised. If, at the end of this time, diplopia and ocular discomfort persist, then surgical adjustment of the visual axes may be indicated. An exception to this rule is in the case of the so-called 'blow-out' fracture of the orbit, in which one or more ocular muscles become incarcerated between the lips of a fracture of the orbital floor. The immediate result of this injury is limitation of movement of the injured eye, particularly in upward gaze. The floor of the orbit must be explored as soon as possible and the trapped muscle freed from the floor of the orbit.

Occasionally, injuries to the posterior part of the orbit, particularly those involving the optic canal, lead to haematoma formation in the sheaths of the optic nerve and sometimes to bruising of the nerve itself. The finding of an inactive or sluggish pupil in a case of this kind will raise the suspicion of optic nerve damage. If the patient can co-operate in simple visual field testing, field loss is often found and this commonly involves either the upper or lower field of vision. The result of such an injury is immediate visual defect which may recover during the subsequent few days or weeks, but occasionally is permanent, in which event it is followed by optic atrophy, and the optic nerve head becomes pale three or four weeks after the injury.

Part Two: The Eye in Diseases of Other Parts

CHAPTER THIRTEEN

The Eye in Cardiovascular Disease

In the eye we are able to see the vascular system of the retina, as it is in life, and examination of the retinal blood vessels gives a guide to the condition of vessels of comparable size elsewhere in the body, especially those within the skull.

As it enters the eye at the optic nerve head the central retinal artery divides into four main branches; these spread outwards, dividing as they go, into the four quadrants of the fundus. Accompanying these arteries are the tributaries of the central retinal vein and the veins are frequently crossed by their accompanying arteries and also cross them. In the healthy vascular tree the arteries become steadily less in calibre as they are followed towards the periphery. They are narrower than their corresponding veins in the proportion of about one to two and there is no interference with the direction of either the vein or the artery where a crossing takes place. The veins are somewhat more variable in calibre and are appreciably more tortuous in their arrangement. There is, on the temporal side of the optic nerve head, an area which is relatively avascular. This is the macular area and it contains a central depression, the fovea centralis (Fig. 6).

In ophthalmological and medical literature the fundus changes in various types of vascular disease are described as being different, depending upon whether arteriosclerosis occurs with or without hypertension, whether or not there is renal involvement, and so on. For these more technical descriptions the reader is referred to larger textbooks dealing in detail with ophthalmological appearances. Here it is proposed to describe only the type of vascular condition which may be visible with the ophthalmoscope and to point to some of the varieties of vessel change which are to be seen.

Probably the first sign of retinal arterial disease is a change in calibre of the arteries. This change may be transient at first and the position of the narrowed segments of the vessel may vary from day to day. This indicates that the calibre change is of the nature of a vasospasm, and it is seen characteristically in toxaemia of pregnancy, when one of the first manifestations of involvement of the retinal vessels is transient arterial spasm. At a later stage of the same process, the retinal arteries become increasingly sclerotic and variations in calibre are now fixed. At the same time as these calibre changes are becoming visible, there are also to be seen variations in the arteriovenous crossings. As has been said, there is normally no interference with artery or vein as they cross one another. In arteriosclerosis, however, the artery, being bound up in a common sheath with the vein, interferes with the underlying vein in a number of ways. The first of these is by a change in the direction of the crossing. The vein, instead of crossing the artery obliquely, now turns sharply in order to cross the sclerotic artery at a right angle. At the same time the blood flow in the vein is impeded and the vein appears to be nipped by the overlying artery, while the segment of the vein distal to the crossing becomes turgid and dilated. It is at such a crossing that venous thrombosis may occur in the retina (Fig. 44).

These changes are practically universal in the elderly and are merely manifestations of the ageing vascular tree. If they become of grosser degree, they are associated with haemorrhages and exudates. In pure arteriosclerosis the exudates are scattered in small groups about the posterior pole of the eye, where they are shiny and rather hard-looking yellowish dots, tending to become confluent. The haemorrhages are most commonly in the surface layers of the retina; they usually appear as 'streaks of blood' sometimes becoming large and filling a large area of the fundus but often simply just a trace of colour in relation to a small vein. In hypertension the haemorrhages are larger and more numerous and the exudates softer and less well-defined. Especially is this the case if there is renal involvement. In the more severe types of vascular disease, frank retinal oedema occurs, manifesting itself principally by oedema around the optic nerve head. The margins of the optic disc become less well-defined and the disc is frankly swollen. The importance of such a finding is that it is necessary for this condition to be distinguished from the papilloedema due to raised intracranial tension. The oedema tends to spread towards the macular region and may produce a characteristic fan-shaped distribution of exudate between the optic disc and the macula, the so-called macular star.

Subjectively, the degree of visual impairment produced by retinal vascular disease is not always easily related to the ophthalmoscopic appearances, for some patients who have very little in the way of abnormality to be seen with the ophthalmoscope complain of considerable impairment, while other patients with

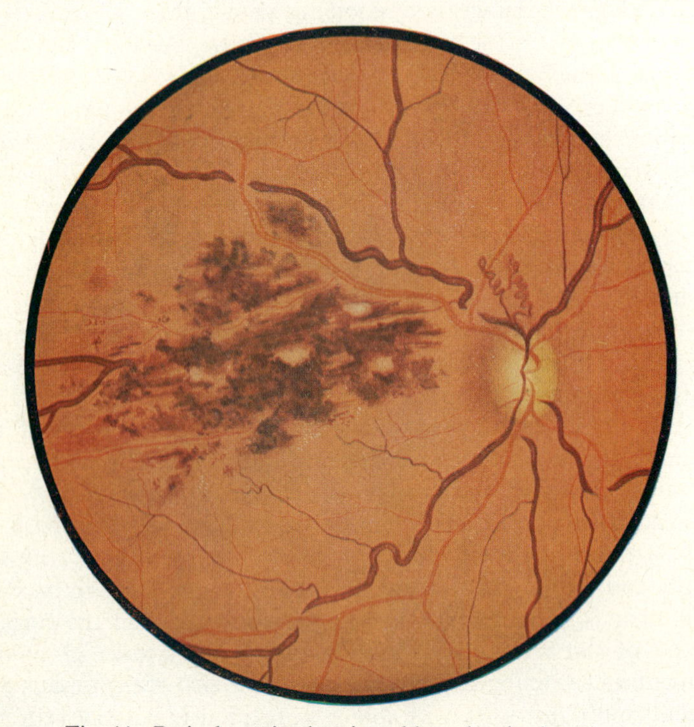

Fig. 44 Retinal arteriosclerosis and branch vein thrombosis.

advanced degrees of 'vascular retinopathy' retain remarkably good vision. The ultimate degree of vision preserved depends upon the extent to which the macula is involved. A small haemorrhage at the macula may destroy central vision, while a patient who has multiple retinal haemorrhages and exudates may retain useful vision, particularly for reading, if the macular area happens to be spared.

Even before there are any haemorrhages and exudates in the retina a patient suffering from arteriosclerosis may complain of transient interferences with vision, often of the nature of flashes of light, sparks or occasionally transient 'black-outs'. These phenomena are of the nature of a vascular spasm causing transient interference with retinal nutrition.

The so-called macular degenerations which occur in the older age group, and which are discussed under diseases of the retina, have their basis in generalised cardiovascular disease.

It is often a comfort to the patient suffering from severe retinal vascular disease to be told that he will not become blind, for, although central vision may be completely destroyed by retinal haemorrhages and exudates, it is unlikely that complete loss of function will follow and peripheral vision is often preserved.

It should be remembered that cardiovascular disease may produce visual impairment in other ways than by direct damage to the retina, and interference with the visual field and gross incapacity may result from interference with the blood supply of the visual pathway within the skull.

Ocular Complications of Arteriosclerosis

1. VENOUS THROMBOSIS. Retinal vein thrombosis is prone to occur at sites where a vein is crossed by a sclerotic artery, in which case the thrombosis is partial and simply involves a branch of the main vein. The commonest site for this to occur is in the upper temporal vein. On the other hand, thrombosis may take place actually at the optic nerve head and the whole retinal venous system may become obstructed.

The patient will complain of visual defect and the severity of this will depend upon the degree of involvement of the retina. If this involvement is purely sectorial and does not involve the macular region, there may be very little complaint, though an intelligent patient will often notice a deficiency in that part of his visual field related to the lesion. On the other hand, in a total venous thrombosis gross visual impairment occurs and the vision may be reduced to the mere perception of light.

With the ophthalmoscope, the veins in the affected area are seen to be grossly distended and tortuous, are considerably nipped by their accompanying arteries where crossings take place, and are the source of multiple retinal haemorrhages (Fig. 44). These haemorrhages are so numerous as to fill the entire fundus in cases of total venous obstruction and they are later accompanied by fluffy ill-defined exudates.

It is never possible to foretell the degree of visual impairment which will result from venous thrombosis. In some cases, particularly in those occurring in the relatively young patient and perhaps due to some other cause than systemic arteriosclerosis, recovery is often good, but on the whole, in the elderly, an eye which sustains a total venous thrombosis never recovers useful function. The same applies to an elderly patient who sustains a branch vein thrombosis, if the branch involved is one responsible for supplying the macular area.

A proportion of patients who sustain a total venous thrombosis in the retina develop in the subsequent months a peculiarly intractable variety of glaucoma, characterised by haemorrhages and new vessel formation in the iris, and by an extremely poor response to surgery (thrombotic glaucoma). The pain resulting from this may be so severe as to lead to the need to remove the eye.

There is another possible relationship between retinal vein thrombosis and glaucoma, and this is with chronic simple glaucoma. It is recognised that retinal vein thrombosis may be a presenting sign of chronic glaucoma. Presumably the raised intraocular pressure compresses the central retinal vein and leads to obstruction of the blood flow.

2. EMBOLISM OF THE CENTRAL RETINAL ARTERY. This is one of the dramatic occurrences of ophthalmology, for it produces sudden painless total blindness of an eye, often without previous warning, but occasionally following attacks which are now considered to be of transient vasospasm. The patient suddenly notices that he is not seeing with one eye and the vision may be reduced to the absence of perception of light. It is natural for the patient to delay taking advice in this condition, for he hopes that spontaneous return of vision will take place, perhaps during the subsequent day or two. But it is unlikely that useful vision can be restored after more than perhaps an hour or two of retinal ischaemia.

Examination of the eye shows a dilated pupil which fails to react to a direct light or reacts poorly while the consensual response is retained. The ophthalmoscope shows an abnormal pallor of the whole of the posterior part of the globe. Through this pallor the macular area shines brightly as a red spot, due to the fact that the retina is thin at the macula and the underlying choroid is, therefore, seen through it. The general pallor and greyness of the remainder of the fundus is due to retinal oedema.

After a few weeks, as the oedema subsides, the fundus regains its normal appearance from the point of view of its colour. As the nervous tissues of the retina have failed to survive the period of ischaemia, however, the optic nerve becomes atrophic and pallor of the optic disc develops. Visual recovery does not, as a rule, take place.

There is some evidence that treatment of this condition in the early stages will produce retinal vascular dilatation and preserve some eyes which might otherwise be lost. This treatment consists of the retrobulbar injection of vasodilators, and the administration of Priscol and nicotinic acid by mouth, but the chances of success are poor.

Obstruction of the central retinal artery may also occur in embolic conditions, such as valvular disease of the heart, and so it may be seen in a relatively young patient. In these cases, one or more glistening emboli may be seen in the retinal vascular tree.

A similar condition, sometimes bilateral, occurs in temporal arteritis. The patient is invariably elderly, shows a raised E.S.R. and systemic disturbance, and may or may not have tender, palpable temporal arteries. Temporal arteritis is treated with systemic steroids and the pain and headache are treated by excision of a segment of the temporal artery. Not only does this relieve the pain but it allows histological confirmation of the diagnosis. This condition is a particularly important one. We can go so far as to say that any elderly patient, complaining of vague malaise, who shows an erythrocyte sedimentation rate of 60 mm or more should have a biopsy of the temporal artery on diagnostic grounds. If the characteristic histological changes are found, treatment with prednisolone holds out good prospect of saving the vision of one or both eyes. An initial dosage of 40 mg prednisolone daily is gradually reduced as the E.S.R. returns to normal, until a small maintenance dose is reached, and this is continued for some months. Temporal arteritis with visual involvement must be treated as a matter of great urgency and treatment must not be withheld pending histological confirmation of the diagnosis. Any patient presenting as a case of sudden painless blindness of one eye and who is found to have a raised E.S.R. must be given systemic prednisolone at once in the dose suggested. It is only in this way that one can be hopeful of preventing blindness of the second eye, which occurs in more than 50 per cent of untreated cases.

CHAPTER FOURTEEN

The Eye in Intracranial Disease

It is with the neurologist and the neurosurgeon that the ophthalmologist has his closest links, and it is intended to point out the various ways in which the eye may be involved in intracranial disorders.

The first importance of the eye in diseases within the skull is in their diagnosis, for lesions within the skull produce interference with visual function, the nature of which may be helpful in the localisation of disease. The condition of the eye, therefore, may be regarded as just one of the clues leading to the building up of the complete picture. On the other hand, diseases of the nervous system not uncommonly present as ophthalmological problems, and it is only later that the true state of affairs is appreciated. The types of disease likely to produce ophthalmological manifestations include tumours, inflammatory lesions and vascular conditions, and injuries.

Papilloedema

Probably the most important single sign of raised intracranial tension, papilloedema, is, in its early stages, extremely difficult to diagnose. The picture may change rapidly and, in doubtful cases, there is nothing more valuable than a repeated examination, after an interval of a few hours. The fully developed condition is unlikely to be mistaken, and it consists of vascular engorgement in the retina, this engorgement being primarily of the veins, which become swollen and convoluted and show, close to the margins of the optic disc, small streaky haemorrhages in the superficial layers of the retina. The next sign of papilloedema is blurring of the margins of the optic disc, and one should remember that the optic disc also appears blurred under conditions of high refractive error, and it is necessary to focus the optic nerve head by using the various lenses available in the ophthalmoscope to exclude this diagnostic difficulty. In addition to the edges of the disc appearing blurred, it is seen that the disc itself is swollen. This is manifest by the appearance of the retinal vessels, as they run over the edge of the disc, where they can be seen to be raised above the general level of the retina. The central pit or cup, which is normally present in the optic nerve head, becomes obscured by oedema and the whole optic disc becomes uniformly elevated (Fig. 45). Papilloedema is usually bilateral, though often more developed on one side than on the other. It is not considered that the degree

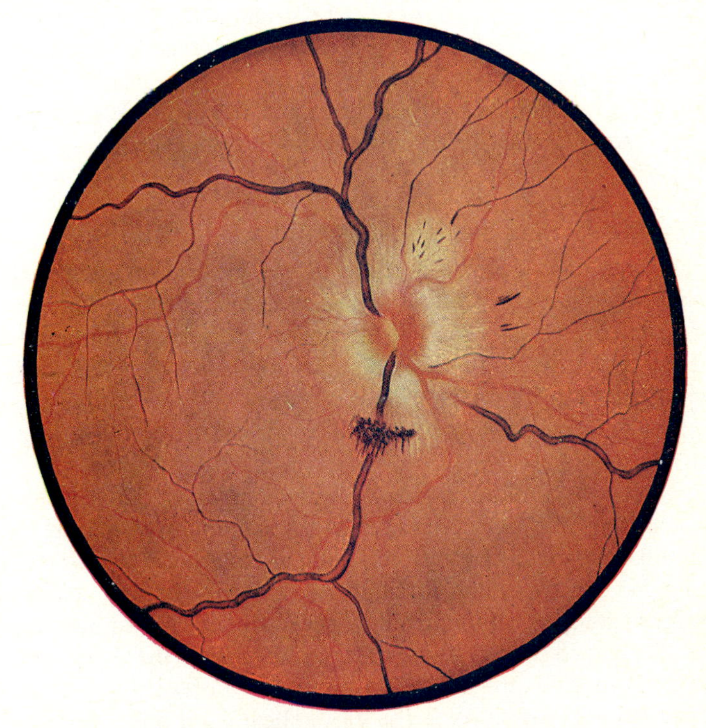

Fig. 45 Papilloedema in intracranial tumour.

of development on one side as compared with the other has any particular localising value, save in certain cases of frontal tumour in which the tumour presses on one optic nerve and produces papilloedema on the opposite side.

In uncomplicated papilloedema, that is to say, papilloedema due to a lesion which is not interfering with the visual pathway, vision is good, although it is rather characteristic for there to be transient attacks of blurring. Examination of the visual field shows some degree of concentric contraction, and enlargement of the blind spot commensurate with the degree of optic disc swelling. This preservation of central vision may help one to distinguish papilloedema from acute retrobulbar neuritis. Acute retrobulbar neuritis, when it occurs close behind the optic nerve head, may give rise to considerable swelling of the optic nerve; although not usually associated with venous congestion or with haemorrhages. In retrobulbar neuritis, however, the visual defect is considerable and vision may be reduced to the perception of hand movements, or even the mere perception of light, by the presence of a dense central scotomatous defect in the visual field. There is also, in inflammatory optic nerve lesions, a sluggish direct pupillary reaction.

Interference with the Visual Pathway

The eye may serve as a guide to the location of intracranial disease by the nature of interference with the visual field, and rough testing of the visual field can be carried out without any special apparatus by the confrontation method which has already been described. Carefully applied, this test will reveal the grosser degrees of visual field loss and will be of considerable help in the diagnosis.

The lesions producing visual field changes are multiple, but the characteristic defects that they give rise to are well defined. If the lesion lies in front of the optic chiasma the defect is strictly uniocular, and we usually find interference with the pupil function on the same side, although the consensual reaction to light from the other eye is intact, owing to the integrity of the third cranial nerve. A tumour pressing upon optic nerve or an injury leading to haematoma formation in the nerve sheath will often lead to the development of optic atrophy, which appears some three or four weeks after the incident.

At the chiasma the characteristic field change is bitemporal hemianopia, due to interference mainly with the crossing fibres in the chiasma—that is, the fibres derived from the nasal halves of each retina. Although this characteristic picture is sometimes seen in pituitary tumours, it is usual for the lesion to be much more advanced on one side than it is on the other (Fig. 46A).

Behind the chiasma any interference with the visual pathway will necessarily affect both eyes. The right optic tract, optic radiation and visual cortex contain fibres emanating from the

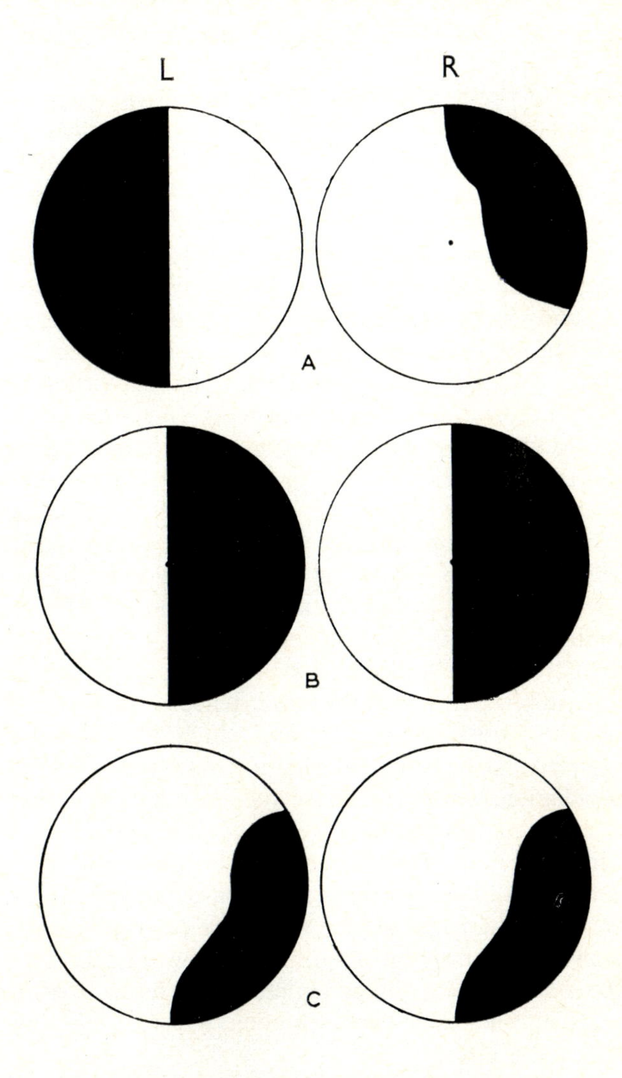

Fig. 46 Visual field changes, charted as seen by the patient.

(A) Bitemporal hemianopia. Pituitary lesion. Complete in the left eye. Incomplete in right.

(B) Homonymous hemianopia. Complete lesion of left visual pathway. Right-sided hemianopia.

(C) Incomplete homonymous hemianopia.

right halves of each retina and, therefore, related to the left visual field, while the left visual pathway behind the chiasma is related to the right half of the field of vision. Any lesion behind the chiasma, therefore, produces what is known as a homonymous hemianopia, one which lies in the same half of the field of vision in each eye. There are, however, certain characteristics which distinguish those lesions lying far forward from those that lie towards the posterior part of the pathway, and these features depend upon the similarity in shape and extent between the visual field loss of the right eye as compared with the left. The farther back in the visual pathway the lesion occurs, the more similar become the two visual field defects. They become in fact more completely congruous, and match one another almost exactly when the lesion occurs far back in the pathway (Fig. 46B, C). There is another feature of the lesions affecting the optic radiation which is of importance and that depends upon the fact that the upper part of the radiation is related to the lower part of the visual field, for it carries fibres arising from the upper retina of the two sides, and the radiational fibres spread out into two well-defined bundles as they sweep backwards through the hemisphere in relation to the descending horn of the lateral ventricle. A lesion approaching the radiation from below affects first those fibres carrying impulses from the lower retina and, therefore, tends to produce an upper visual field defect, while a lesion which is approaching the radiation from above, e.g. a parietal tumour, presses first upon the upper radiational fibres and so produces a lower visual field lesion (Fig. 46D).

A similar state of affairs arises with regard to the distribution of the fibres in the visual cortex. In the occipital cortex the fibres subserving the upper retina lie above the calcarine fissure, while fibres serving the lower retina terminate below the calcarine fissure, and the more centrally placed regions of the retina, that is to say, the macula, find their representation towards the tip of the posterior pole (Fig. 46E).

It is on anatomical facts such as have been outlined that an interpretation of visual field defects depends, and, although detailed analysis may depend upon accurate perimetry with elaborate apparatus, the majority of field defects when fully developed can be distinguished by simple confrontation methods.

Movements of the Eye and Sensory Nerve Lesions

In addition to interference with the visual field, lesions within the skull may produce disturbance of ocular movements or sensory changes.

Lesions of the oculomotor nerves often give rise to diplopia, and examination of the eye movements shows a limitation of

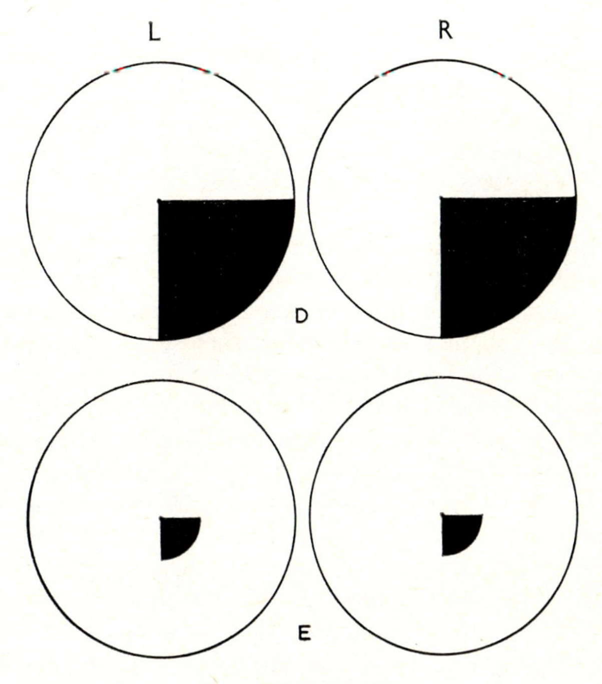

Fig. 46 (continued)

(D) Quadrantic homonymous hemianopia. Lesion in upper part of left visual pathway.

(E) Central quadrantic homonymous hemianopia. Lesion at tip of visual cortex, on the left side, above the calcarine fissure.

movement in one or other direction. From what has already been said with regard to the differential diagnosis of paralytic squint, it will be seen that the diagnosis depends upon the knowledge of the action of the various extraocular muscles and the analysis of the nature of the diplopia. Ptosis may occur in lesions affecting the third cranial nerve. In addition to disturbances of the nerve trunks, we sometimes see disturbances of the supranuclear pathways, when the paralysis is one rather of movement than of specific ocular muscles. A patient suffering from such a condition will be unable to look voluntarily upwards or downwards or to right or left, while special testing may show that certain reflex movements are preserved and that the muscles and their connections are, therefore, intact, thus placing the lesion above the nuclei in the midbrain.

DISORDERS OF THE PUPIL

In any abnormality of the pupil function, close examination should be made to exclude local disease, for in iritis the pupil becomes adherent to the lens and may be mechanically bound down on this account. The exclusion of this type of condition will help to avoid a number of 'false alarms'.

The normal pupil reacts briskly and sustains its contraction when light is shone into the eye or when light is shone into the opposite eye. The simplest way of examining pupil function is to seat the patient facing a window and to cover both eyes with the examiner's hands. Each eye is then uncovered in turn and the direct pupil reaction examined. While holding one eye shaded, the opposite eye is now uncovered and the pupil of the shaded eye can be watched. This allows inspection of the consensual reaction which should be brisk and which depends upon the bilateral transference of impulses from the optic tract to the third nerve nucleus in the midbrain. The reaction to convergence should next be tested by the patient looking first into the distance and then to the examiner's finger held six inches from the patient's face and in the midline. There should be a brisk reaction of the pupil to the movement of convergence.

Some of the commoner and more characteristic disorders of pupil function are described below.

ARGYLL ROBERTSON PUPIL. This is characteristic of cerebral meningovascular syphilis and is usually bilateral. A similar disturbance of the pupil is sometimes seen in diabetes. The characteristic pupil is small, irregular in outline, and reacts not at all to light, either directly or consensually. It does, however, react to accommodation and convergence. It is considered that the lesion lies in the pupil pathway between the optic tract, where the pupil fibres leave the visual fibres to enter the roof of the midbrain, and the third nerve nucleus.

THE TONIC PUPIL, OR THE HOLMES-ADIE PUPIL. The Adie pupil is semi-dilated and on first examination appears to be unreactive to light. If, however, the patient is exposed to the bright light for some minutes, the pupil contracts slowly and dilates again equally slowly on return to the dark. It occurs unilaterally in young people and is associated with absence of the deep reflexes in the limbs. The condition is without special significance and is unassociated with any neurological disorder.

RETROBULBAR NEURITIS. In acute retrobulbar neuritis the pupil on the affected side reacts slowly to direct light and the contraction is poorly sustained. On the other hand, contraction of the pupil to consensual stimulation is intact.

HORNER'S SYNDROME. The complete Horner's syndrome comprises a small pupil, ptosis, and diminished sweating on the affected side of the face, together with an apparently somewhat sunken eye. The condition is sometimes congenital in origin but, if it is due to an active process, may be caused by any disease or injury affecting the sympathetic pathways. Among the possible conditions are cervical rib, arthritis of the cervical spine, apical lesions in the chest and tumours in the neck and also injuries to the upper part of the thorax and the cervical region.

THE BLIND EYE. In complete lesions of the optic nerve, such as occur after contusion injuries or destruction of the nerve by any pathological process, perception of light is lost and with it the direct reaction of the pupil to light. The pupil of the opposite eye also fails to react when light is shone into the affected eye. If, however, the third nerve and its connections are intact on the affected side, then the pupil of the blind eye will contract when light shines into its other eye. This is the consensual reaction.

ACCIDENTAL MYDRIASIS. It must be remembered that the commonest cause of sudden unilateral dilatation of the pupil without other signs is accidental exposure to mydriatic drugs. This may be due to the use of eye drops or ointment ordered for another patient, or to contamination of fingers and transfer of the drug to the eye. This latter type of accidental medication may occur in nurses working with atropine, in mothers who have been putting atropine into a child's eye, or to the use of lotions or liniments containing belladonna.

OPTIC ATROPHY

Optic atrophy implies destruction of the nervous elements in the optic nerve. Descriptively, the terms primary and secondary have been applied to optic atrophy, but these terms are based purely on ophthalmoscopic appearances and convey no information as to the aetiology of the condition. They tend, therefore, to be misleading, particularly when it is realised that the same process occurring at different parts of the nerve may give rise in one instance to primary, and in another to secondary atrophy. Secondary optic atrophy follows some inflammation or swelling of the nerve and is seen after the subsidence of papilloedema and of an attack of retrobulbar neuritis close to the optic nerve head. Primary atrophy on the other hand follows some lesion in the optic nerve which does not produce oedema and swelling of the optic nerve head. Among such conditions are contusions of the nerve, perhaps due to fractures at the apex of the orbit, certain toxic processes interfering with the nutrition of the ganglion cells of the retina, and also glaucoma, the destruction of the nerve head by which disease results characteristically in increasing pallor of the nerve head (Fig. 40).

The diagnostic feature of optic atrophy is pallor of the nerve, and the nature of the resultant visual defect will depend largely upon the type of disease which leads to the atrophy. Thus, in retrobulbar neuritis and allied conditions, there will be a visual defect in the central area of the visual field, while in conditions such as tabes dorsalis, which is the most typical cause of primary optic atrophy, the lesion will most likely be a gradual contraction of the visual field in the periphery with some falling off of central visual acuity.

One particular variety of optic atrophy which should be mentioned is that which follows haemorrhage, such as haematemesis or uterine haemorrhage. This post-haemorrhagic optic atrophy does not seem to occur after haemorrhage due to trauma.

CHAPTER FIFTEEN

The Eye in Endocrine and Metabolic Disease

DIABETES

Diabetes leads to iritis, and it was considered in the past that diabetic iritis was a specific condition differing from other varieties of intraocular inflammation. It is now believed that the iritis in diabetics is simply a manifestation of the well-known fact that diabetic patients are more prone to inflammatory conditions of one sort and another than are non-diabetic patients, and that the diabetic iritis is simply an ordinary iritis occurring in a diabetic patient.

Treatment, therefore, is no different from that of iritis in other cases and it consists of mydriatics and steroids locally, heat and some analgesics for the relief of pain.

Another thing which may occur in diabetes is interference and change in the refractive state of the eye. Not only may the actual basic refraction of the eye alter, and alter quite quickly if the diabetic state is unstable, but there may be frank paralysis of the accommodation. This paralysis occurs in diabetic patients who have lately begun to use insulin and passes off during the subsequent few weeks. It is manifest by an inability to focus for close objects.

Interference with pupil function has been described and the typical Argyll Robertson pupil sometimes occurs.

The main importance of diabetes from the ocular point of view, however, lies in the effect that it has upon the retinal vessels, in the production of so-called diabetic retinopathy. This is an important and sometimes blinding complication of diabetes.

The onset of the retinopathy seems to depend partly upon the age of the patient and partly upon the length of the time for which the diabetes has been present. The younger the patient, the longer in the course of illness it seems to take for the retinopathy to develop. An older patient may apparently develop diabetic changes in the fundus after having had diabetes for a relatively short time. A disappointing feature of the condition is that control of the diabetes is not certain to prevent the development of retinopathy, or to prevent its further progression once it has commenced.

The first sign of diabetic retinopathy is the appearance of small round haemorrhages or aneurysms at the posterior pole of the eye, and some generalised venous dilatation. There is doubt as to whether all the small red spots which are seen in the diabetic fundus are haemorrhages or aneurysms, and the difference between

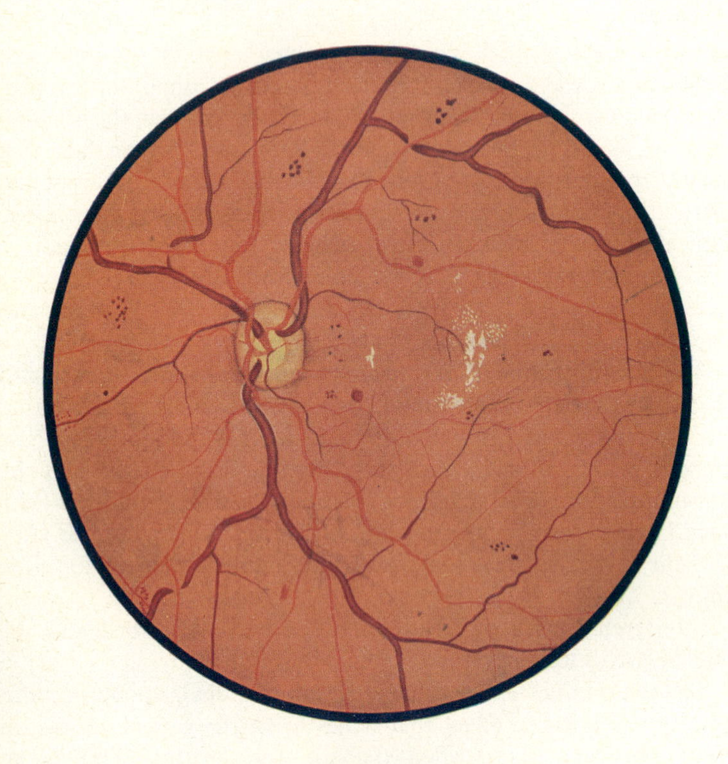

Fig. 47 Diabetic retinopathy.

them is not appreciable ophthalmoscopically. Histologically it is known that diabetes leads to the formation of small venous aneurysms in the retina, and it is these aneurysmal dilatations which may be first seen and which look like deep round haemorrhages. Later, larger haemorrhages may occur and burst through into the vitreous with the production of vitreous haemorrhages and consequent organisation by fibrous tissue. In addition to the haemorrhages there are exudates, characteristically small, round and shiny. The exudates, of creamy-yellow colour, become confluent and are mostly situated in the region around the macula. Depending upon the degree of involvement of the actual macular region so does the patient's vision become defective, but gradual advance of the condition leads inevitably to visual failure (Fig. 47). Although the general rule still holds good, that diabetic retinopathy usually progresses inexorably to produce gross visual defect, there is increasing evidence that some control is possible in the occasional case. One form of treatment depends on the use of substances (e.g. Atromid) which influence the level of the blood lipids and may reduce the formation of diabetic exudates in the retina. The other possibility, in a younger patient with a florid, haemorrhagic type of retinopathy, is that pituitary ablation may delay the onset of blindness. This very radical treatment is only considered in desperate cases and with very strict criteria of case selection. Attempts are also being made to contain haemorrhagic retinopathy by direct coagulation of abnormal vessels in the retina, either by white-light photocoagulation or laser. It is hoped that the incidence of haemorrhage into the vitreous may thus be reduced.

The diabetic patient is also liable to develop senile cataract at a slightly earlier age than does his non-diabetic fellow, and patients presenting for cataract surgery are routinely examined for diabetes: if present, the condition may then be treated before the patient is operated upon. A rarer way in which the diabetes may affect the lens is in the production of what is known as true diabetic cataract, a cataract occurring in a young patient suffering from severe diabetes and due to metabolic changes in the intraocular fluids. In the early stages such a cataract is reversible, if the diabetes can be brought under proper control, but later the opacities become fixed, and the question of surgical treatment may arise.

Thyroid Disease

The patient suffering from thyroid disease may arrive in the eye department as a case of exophthalmos; or there may be difficulty with ocular movements, or recurrent ocular irritation due to imperfect closure of the eye. The diagnosis of a well-developed case of exophthalmos is not usually difficult but the earlier signs are sometimes difficult to interpret.

When the eye is in the normal position, the lower lid crosses the junction between the cornea and sclera below, while the upper lid covers the upper one-third of the cornea. Exophthalmos leads to widening of this aperture, partly by increased prominence or protrusion of the eyeball and partly by retraction of the upper lid, perhaps due to sympathetic overactivity (Fig. 48). If the protrusion becomes considerable, then the question of protecting the eye from damage by exposure becomes the main part of the ophthalmological problem. The cornea does not readily tolerate exposure to the air and consequent drying. Drying leads to breaking down of the corneal epithelium and to infection. This exposure keratitis

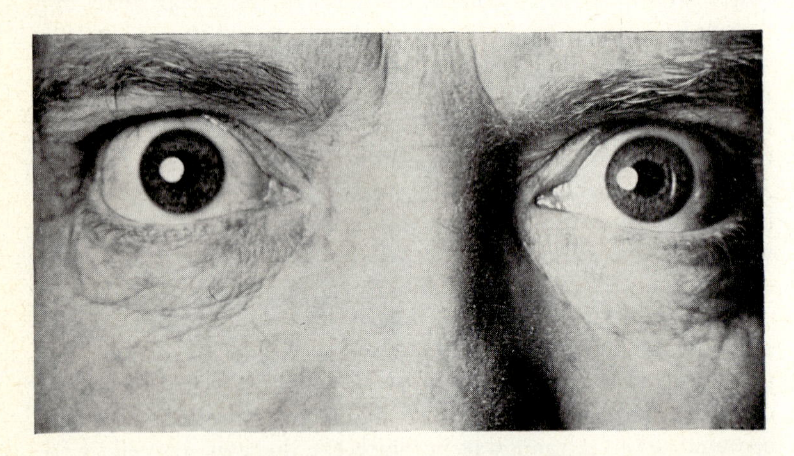

Fig. 48 Lid retraction in thyrotoxicosis.

is seen in the lower third of the cornea where the exposure first occurs and may, if untreated, lead to corneal opacity. Exophthalmos due to thyroid disease is, of course, not the only cause of exposure keratitis, which may result from proptosis due to orbital tumour. It may occur, too, in severe debilitating conditions in which the normal blinking movements of the eyes are absent, or in the presence of facial palsy and wounds involving the eyelids. In all these conditions, the responsibility of the medical attendant is to prevent desiccation of the cornea. This is best done, as a temporary measure, by the provision of abundant lubrication. Some antiseptic ointment, covered by a well-greased pad, will usually prevent corneal damage. In some cases strapping of the eyelids at night to ensure closure during sleep may be useful. If the condition is likely to be a long-standing one and if these measures are not sufficient, then tarsorrhaphy, that is, the production of an adhesion between upper and lower lids in order to reduce the size of the palpebral aperture, may be required. This adhesion may be re-opened at a later date, if the condition originally demanding it resolves.

The problem of aetiology of exophthalmos in thyroid disease is a complicated one, and it is now well established that there are two varieties of the condition. The first is the traditional thyrotoxic exophthalmos, which occurs in hyperthyroidism and which is usually relieved by reduction of thyroid activity. The second variety is characterised by more fully developed ocular muscle involvement and by the presence of considerable orbital oedema. This is not due primarily to thyroid disease, for it may occur after thyroidectomy and in the presence of a normal or low basal metabolic rate, but is due to other hormonal influences acting upon the orbital tissues. Whether these other influences arise in the pituitary (and the pituitary thyrotropic hormone has been implicated in this condition) or whether they arise from the adrenal glands is not certain, but what is established is that there are other hormonal factors which can play upon the orbital tissues, with the production of oculomotor paralysis and exophthalmos. This latter condition is known as exophthalmic ophthalmoplegia.

The responsibility of the eye doctor remains the same, however, under all circumstances, and it is to protect the eye from damage by exposure until the exophthalmos can be reduced by treatment. There is increasing certainty that exophthalmic ophthalmoplegia, at least, is a self-limiting disease, and if we can maintain the patient's ocular condition we may find that the exophthalmos diminishes during the course of months or sometimes years. If treatment directed at the endocrine condition fails to control the exophthalmos, and it is considered that a risk of losing the eye is developing on account either of stretching of the optic nerve or exposure of the cornea and consequent keratitis, then the problem of an actual attack on the exophthalmos itself becomes an important one. The choice lies between orbital decompression, an operation performed through the roof of the orbit by the neuro-surgeon, and radiation of the orbital tissues. This latter produces shrinking of the densely infiltrated orbital fat and extraocular muscles and reduction in the degree of exophthalmos; although this reduction takes several weeks to reach its maximum effect. On the whole, it is considered that radiotherapy is the most useful line of treatment, although, as has been said, the exophthalmos tends to be self-limiting in nature and temporisation may be all that is required. Once the condition has become static, surgical treatment may be needed. One reason for this is cosmetic, due to the staring appearance of the prominent eyes. In this instance, a lateral tarsorrhaphy to reduce the size of the palpebral fissure is often of benefit. A second possible need for treatment is in the management of diplopia following ocular palsy in exophthalmic ophthalmoplegia. This is carried out on the general lines of treatment for paralytic squint (q.v.).

Rheumatic Disease

Interest in rheumatic disease has been increasing during the past years, and the eye appears to play a part in certain of these conditions.

In ankylosing spondylitis, iritis is a common presenting symptom, and patients suffering from iritis are always questioned and X-rayed with regard to the possibility of disease of the lumbar vertebrae.

Episcleritis is reputed to be related to rheumatism. The association here is not so clear, but it seems that some patients suffering from episcleritis improve when treated empirically with salicylates by mouth and some, though not all, admit to having a rheumatic history.

One important relationship between the eye and rheumatic disease is in a peculiar variety of keratitis and conjunctivitis which is quite certainly associated with rheumatoid arthritis. This is kerato-conjunctivitis sicca, or Sjögren's disease, characterised by diminished secretion, both of the lacrimal and salivary glands, by dryness of the cornea and the formation of epithelial filaments on its surface. The patient complains of vague irritation and attacks of redness of the eyes and is often suffering from well-developed rheumatoid arthritis. Examination of the quantity of the tears secreted, which can be carried out by placing a narrow strip of filter paper between the closed lids, and estimating the rate at which it is wetted by the tears, shows that there is gross lack of secretion of tears when compared with normal. (The normal lacrimal secretion will moisten a strip of filter paper five millimetres wide for a distance of some 15 mm in five minutes.) The response to treatment is variable, but some of these patients are made much more comfortable by the use of artificial tear fluid, and one of the most popular tear replacement fluids is methyl cellulose in a strength of 0·6 or 1 per cent. As an alternative there is a simple Ringer's solution, although this does not remain in the conjunctiva for such a long time as does methyl cellulose. Steroids have been tried in a number of cases and give variable results, some patients saying that they are more comfortable when using steroids locally. A further measure which helps sufferers from Sjögren's syndrome is surgical occlusion of the lacrimal drainage puncta. This produces obstruction to the outflow of the tears and so preserves the tears which are being secreted and enables the patient to make the best use of what tear fluid there is available.

Other Conditions

Few other ocular conditions are certainly ascribable to endocrine and metabolic disorders, though there are some which, presumably by interference with the nutrition of the lens of the eye, produce cataract. Among these should be mentioned mongolism, myotonia dystrophica and galactosaemia. It should also be remembered that toxaemia of pregnancy, the aetiology of which is still uncertain, may be associated with vascular changes within the eye comparable with those seen in malignant hypertension.

CHAPTER SIXTEEN

The Eye in Diseases of Anatomically Related Parts

(EXCLUDING INTRACRANIAL DISEASE)

DISEASES OF THE SKIN

The eyes and eyelids show consecutive inflammation in certain diseases of the skin, among which are impetigo, seborrhoea and acne rosacea.

SEBORRHOEA CAPITIS

In seborrhoea capitis chronic squamous blepharitis and conjunctivitis are so common as to be almost a rule. The patient complains of chronic irritation of the eyes with some redness and occasional stickiness in the mornings. There are to be seen pale scales, aggregated around the roots of the eyelashes and a mild degree of coincident conjunctivitis. Without treatment of the scalp, local treatment to the eyelids can produce only temporary improvement. Shampoos containing Selenium are very effective in the treatment of dandruff. The eyelid condition will then resolve rapidly if, after removal of the scales by warm bathing, sulphacetamide ointment is firmly massaged into the roots of the eyelashes at night. As part of the redness and irritation of the eyelids represent a reaction to the presence of scales on the eyelids, steroids may be useful, in the form of either drops or ointment. The treatment to the scalp must be continued, even though the condition seems to be under control, for remissions are very frequent; the shampoo should continue to be used once a week or more frequently as the need arises. Sufferers from seborrhoeic blepharitis occasionally develop crops of styes. This should make us aware of the possibility that the scalp may also be involved.

IMPETIGO

In impetigo a frank ulcerative blepharitis may appear; it demands treatment on general lines. There should be careful removal of the crusts from the roots of the eyelashes and application of antiseptic ointment, usually chloramphenicol or framycetin, firmly massaged into the eyelid margins. This, together with treatment of the skin as a whole and associated with attention to the patient's health and nutrition, will always produce improvement.

Acne Rosacea

This is perhaps the skin disease with which the ophthalmologist comes most frequently into contact. A considerable proportion of patients suffering from acne rosacea develop chronic conjunctivitis and blepharitis, often associated with recurrent meibomian cysts, and they are also prone to a particularly destructive variety of keratitis. This keratitis may occur in the absence of severe skin involvement, but evidence of skin lesions is always to be seen, if carefully looked for. As in the case of the skin, the keratitis is prone to remissions and recurrences, these not necessarily being related to similar phases of the disease in the skin. If effective treatment is not available, a severe vascularising superficial keratitis, almost always bilateral, will arise and will give rise to photophobia, lacrimation and gradually increasing visual failure. It is characteristic of the corneal lesions for them to develop as tongues or patches of opacification spreading in from the corneal margin. The opacity ultimately will affect the pupillary area of the cornea and produce gross visual disability. Apart from the possibility of visual trouble, the patient is made repeatedly uncomfortable by the recurring attacks of irritation and inflammation. Although treatment with mydriatic drugs, such as 1 per cent atropine or 0·25 per cent hyoscine drops, may be used, these are often not effective in controlling recurrent attacks of keratitis. Steroids, on the other hand, in the form of either drops or ointment (depending upon which suits the particular patient best) will almost always keep the disease under control. As each attack is self-limiting, we can provide our patients with a supply of betamethasone or dexamethasone to use when the condition is active and can, therefore, usually prevent further spread of the corneal opacity. It is not in every case that cortisone is helpful, but the majority of cases respond to this treatment. In the occasional unilateral case, where corneal disease is of advanced degree on one side, it is often justifiable to perform a tarsorrhaphy and so produce temporary closure of an eye and so to give the cornea a chance to heal. Not only does this procedure remove the constant irritation of recurrent attacks of photophobia, but it seems to assist the healing process. If severe visual defect is present, corneal grafting may be indicated.

Among other skin conditions which sometimes have to be considered in connection with the eye disease are:

1. *Pediculosis*, in which examination of the eyelashes shows the offending organisms clinging to the hairs. Manual removal of the nits is probably the most effective treatment when combined with treatment to the scalp or other parts of the body involved.

2. *Pemphigus*, in which a condition known as Essential Shrinkage of the Conjunctiva may produce gross scarring of the tissues and sometimes distortion of the lids and opacification of the cornea.

3. *Stevens-Johnson syndrome*, where an acute purulent conjunctivitis is associated with febrile illness, glandular swelling and a tendency to spontaneous resolution. The severe vascular reaction of the conjunctiva may lead to secondary corneal involvement.

4. *Reiter's syndrome*, where a non-specific urethritis is associated with conjunctivitis, iritis, and polyarthritis.

5. *Behcet's syndrome*, a condition in which a destructive bilateral iridocyclitis is associated with buccal ulceration and ulcerative lesions of the genitalia.

EAR, NOSE AND THROAT DISEASE

Diseases of the neighbouring sinuses have to be considered in any inflammatory condition of the orbit. Orbital cellulitis, which shows itself as proptosis and fixity of the globe, associated with fever and chemotic swelling or oedema of the conjunctiva, is frequently secondary to sinusitis. This sinusitis may be in the ethmoids, where the swelling presents mainly in the antero-medial part of the orbit, or it may be farther back, in which case a straightforward proptosis is the rule.

We have also to consider lesions arising in the nose or nasal sinuses in cases of unilateral proptosis, for tumours arising in antrum or ethmoidal air cells may burst through the thin wall separating them from the orbit and give rise to proptosis and displacement of the globe to one side or another. The opinion of an ear, nose and throat surgeon is desirable in any condition of this nature. Not only may tumours spread in this way into the orbit, but diseases of the maxillary antrum are also potential causes of unilateral lacrimal obstruction, due to interference with the naso-lacrimal duct in its lower part, and nasal disease should be excluded as a possible cause of dacryocystitis, if there is any doubt as to the aetiology in a given case.

There is a further way in which neoplasms arising in the nose and throat territory may affect the eye, and that is by a spread upwards through the base of the skull from retropharyngeal tumours. These tumours present ill-defined symptoms and signs often leading to a rather bizarre oculomotor palsy, perhaps associated with sensory involvement in the distribution of the fifth cranial nerve, and may give rise to difficulties in diagnosis.

The relation between diseases of the nasal sinuses and retro-bulbar optic neuritis is an uncertain one. It has been considered that sphenoidal sinusitis could be a cause of retrobulbar neuritis, but the evidence is far from convincing and it is not considered that ear, nose and throat disease is a cause for such a condition.

CHAPTER SEVENTEEN

The Eye in Other Conditions

DISEASES OF PERIPHERAL NERVES

Retrobulbar Neuritis

Unilateral acute retrobulbar neuritis is sometimes a manifestation of disseminated sclerosis, but a number of patients suffering from an isolated attack of retrobulbar neuritis never develop any further manifestations of disease. The condition is commonest in early adult life.

The acute attack is ushered in by unilateral pain behind the eyeball and this is associated with rapid visual failure, often of gross degree. Not only is the eye painful, but so also are movements of the globe and pressure on the eyeball through the closed lids. In the natural course of events these symptoms subside during the subsequent four to eight weeks and it is usual for visual recovery to occur. This recovery may be virtually complete.

Examination in the acute stage shows that the peripheral field of vision is intact and that the visual loss is confined to a central scotoma of varying density, but usually associated with depression of the vision to less than 6/60. The pupil characteristically reacts slowly to direct light and maintains its contraction poorly. The consensual reaction when light is shone into the opposite eye, is brisk. If the inflammatory patch is far back behind the optic nerve head, the optic nerve will appear normal. A patch of inflammation farther forward gives rise to oedema of the nerve head, manifest by blurring of the margins of the disc and some swelling. This swelling is not of the high degree that we associate with papilloedema of raised intracranial tension; it is not associated with venous congestion and there are no haemorrhages. These features, associated with the fact that the vision in papilloedema is characteristically good, allow us to distinguish acute retrobulbar neuritis from swelling of the optic nerve head due to raised intracranial pressure. In the course of a few weeks after an attack of retrobulbar neuritis, pallor of the optic disc develops and this pallor is usually confined to the temporal half of the disc. This is due to the fact that the macular fibres, which are most sensitive to the destructive effects of inflammation in the optic nerve, lie temporally, and it is these fibres which atrophy.

Although acute retrobulbar neuritis is characteristically unilateral, there are certain conditions in which the inflammation of the nerve is bilateral. Among these is retrobulbar neuritis associated with neuromyelitis optica, or Devic's disease, in which there is an associated myelitis leading to paralysis and sensory changes.

Myasthenia Gravis

Although it is commoner in middle age, myasthenia gravis may occur at any age and present as an ophthalmic problem because the patient may be affected by diplopia or by progressive ptosis. The condition has characteristic remissions and relapses and the state of the eyes may vary from day to day and from week to week; diplopia, for example, is often worse towards the end of the day. Examination shows ptosis, commonly bilateral, and this is increased by artificially exercising the levator of the upper lid, which can be done by making the patient follow a finger up and down before the eyes, or, alternatively, by making him look steadily upwards for a minute or so. The diagnosis can be confirmed by the injection of prostigmine 2·5 mg combined with atropine 0·6 mg. This abolishes the ptosis and the diplopia in about fifteen minutes. As an alternative diagnostic test, Tensilon may be given intravenously in a dosage of 10 or 20 mg.

Herpes Zoster

The relation between herpes zoster and corneal disease has been discussed elsewhere. Shingles may also be a cause of oculomotor palsies, and occasionally of retrobulbar neuritis.

TOXINS AND CHEMICAL POISONS

Tobacco

The smoking of heavy tobacco in a pipe over many years produces disturbance of the ganglion cells of the retina and leads to bilateral visual failure. The failure is almost always equal on the two sides and is manifest at first by failure of appreciation of colour, particularly of red, in that part of the visual field lying close to the fixation point. This colour defect not uncommonly leads to the complaint of difficulty in distinguishing between copper and silver coins without feeling for the knurled edge. The condition is sometimes associated with excessive drinking and is more severe in patients who are also diabetic. Whether or not the condition has some association with faulty metabolism of vitamin B is uncertain, but it is a possible explanation. Tobacco amblyopia has to be considered as a possibility in the diagnosis of otherwise unexplained visual failure in elderly men. It is excessively rare in cigarette smokers.

Treatment involves complete abstinence from tobacco, a hardship which some of our old patients endure very badly, and those in whom improvement fails to occur are almost certainly continuing to smoke surreptitiously. If smoking is stopped completely, improvement in vision occurs during the subsequent six or eight weeks. There is some evidence that this improvement is assisted by the administration of hydroxocobalamin (Neocytamen) by injection (1000 μg daily for a week, then twice weekly for six weeks, the frequency of the dose being gradually reduced thereafter). When improvement has reached its maximum the patient may resume smoking, but should confine his consumption to a moderate amount.

Methyl Alcohol

Methyl alcohol is a constituent of various illegal liquors and is extremely toxic. It may lead to rapid unconsciousness and death, but, if the toxic effects are less severe than this, there may be severe visual failure sometimes progressing to total blindness. Sensitivity to the toxic effects of this alcohol seems to vary, but blindness has been reported after the consumption of only one ounce of the alcohol.

Quinine

Some people are particularly sensitive to quinine, and gross contraction of the visual fields has been reported after only a small therapeutic dose. The majority of cases of quinine amblyopia, however, occur after the ingestion of large quantities, often in an attempt to produce abortion and the visual symptoms are accompanied by tinnitus. The visual failure due to quinine comes on rapidly during the first twelve hours after taking the poison and is associated with gross contraction of the visual fields with visible narrowing of the retinal arteries, generalised pallor of the optic fundi, and pupils which react poorly or not at all to light. Gradual improvement occurs during the subsequent few weeks, but some permanent visual loss may remain.

Other Chemicals

Among the organic solvents and industrial poisons which may produce ocular manifestations are carbon bisulphide, lead, and arsenic.

Part Three: Administrative

CHAPTER EIGHTEEN

The Welfare of the Blind

Definition

A person is considered to be 'blind' if he does not possess sufficient vision to be able to perform work for which eyesight is essential. This, of course, does not imply an inability to follow a previous occupation, but an inability to do any work demanding vision. The interpretation of the definition varies somewhat between different local authorities, but, in general, a man whose corrected visual acuity in the distance is less than 3/60 is to be regarded as blind. Exceptions to this rule may arise in the case of patients possessing relatively good central vision, but in whom this is associated with gross visual field loss.

Examination with a view to certification can be arranged through the eye department of the local hospital or by the local health authorities.

If the examining doctor considers that the applicant, although not now blind, is likely to become so within the course of the next four years, he should state this on the form, and the case will then come under the care of the blind welfare authorities.

Blind Children

The best arrangement is probably for a blind child to remain in the home surroundings until school age is reached, providing that the home conditions are good; in this way the sense of being a member of the family may be preserved. Unless parents are more than usually sensible, however, there is a risk either that the blind child will suffer from lack of attention and be led to feel an outcast among his normal brothers and sisters, or that the latter will be neglected through his becoming the centre of attention. Should this prove to be the case, the child may be better off in a residential nursery school. These are to be found throughout the United Kingdom and are known, in England, as Sunshine Homes.

Children may enter a nursery school of this kind from the age of two years.

If the child is to remain at home until normal school age, the Blind Welfare Authorities will arrange for teachers to call at the home to advise the parents about handling their problem. As soon as the child begins to move about the house, bumps and bruises are unavoidable, but he will very soon learn his way about in the familiar surroundings. Toys have to be chosen with care to avoid those with sharp points and edges. It is important that the parents should be able to take little notice of the injuries that are bound to occur at first and to let the child develop his own technique for finding his way about.

When the age of five is reached, the responsibility for providing education for the child is placed, by the various Education Acts, on the local authorities. There are, in each large centre, residential schools for blind children, and it is to one of these that the child must go. Education here is of a general nature and is specially planned for the needs of the blind. At school the child is fitted for either a manual or professional career, to which he goes on leaving school at the age of eighteen.

Difficulties lie in the handling of the partially-sighted child; that is, a child who, though not seeing well enough to benefit from normal education, does not fall within the definition of blindness. Although there are, in large centres, schools for the partially-sighted, these are mostly non-residential. It follows, then, that a partially-sighted child whose home is so placed that he cannot attend as a day-boy at one of the schools for the partially-sighted, can be educated only by entering one of the residential schools for the blind.

Those Who Lose Their Sight in Middle Life

These cases will be those of visual loss due to injury, iridocyclitis, detachment of the retina, and other diseases. (Those who are blinded in war are treated in separate establishments, but the principles are the same.)

There are, of course, only relatively few occupations available to the blind and these are mainly of a manual nature.

The condition of a recently-blinded man needs very careful and considerate management, and every stimulus to persuade him to take the essential step of being willing to enter into some activity should be offered. Although the realisation of the fact of blindness may have come gradually, its impact is none the less severe, and many are unable, by themselves, to get over this blow.

The home teachers of the blind welfare authorities will call on the patient at an early stage and will commence his rehabilitation, but the most important step will be his entry into one of the several Rehabilitation Centres which have been established under the Disabled Persons (Employment) Act (1944). Here, among people in a similar condition to himself he will find training for a new employment and the least possible scope for self-pity. The blind man must be taught to stand on his own feet again and this as soon as possible.

At the Rehabilitation Centre the aptitude of the patient for the various possible jobs will be assessed and he will enter into training. On discharge he will most likely be found employment in a factory where a number of posts are filled by disabled workers. Particularly in small machine shops and in simple repetitive tasks demanding fine tactile sense, the blind worker can often more than hold his own with his sighted neighbour.

A difficulty arises in the case of men whose homes are in the country, far from a centre where employment in a factory might be obtained. To move the patient into lodgings in the town naturally disturbs his family life, which has often been disorganised by the fact of his blindness. On the other hand, the chance of the whole family being able to be moved into the town is small, for there is the difficulty of obtaining accommodation, and there is no priority accorded to these patients when they are trying to find a house.

In the case of patients whose sight is defective, but not so bad as to allow of blind certification, the Ministry of Labour may help. The Disabled Resettlement Officer is able to give special attention and consideration to the selection of jobs for these 'partially-sighted' persons.

The Elderly Blind

In general, the aged blind person cannot be employed although there are exceptions dependent upon the patient's previous employment, his ability, and mental outlook.

The Blind Welfare Authorities can, however, do a great deal to assist in the rehabilitation of the blind person and in the guidance of the family in their approach to the problem. Braille can also be learnt, Home Visitors will call, and introductions can be made to other blind people or to clubs catering specially for the blind.

Summary of the Advantages of Registration as a Blind Person (in addition to those described above)

1. FINANCIAL ASSISTANCE. The 'Blind Pension' is not paid simply because a person becomes blind, but may be available to supplement income if other sources do not produce a certain figure.

2. TRAVELLING. (*a*) Free rides in Corporation Transport in most cities, or half-fare on certain country services.

(*b*) On the railway, if travelling in the course of business, a blind person and a companion may travel for the price of one ticket.

Vouchers entitling the patient to these benefits are issued by the Blind Welfare Authorities.

3. WIRELESS AND TELEVISION. (*a*) Free licence for sound radio. If television is used there is a reduction of £1 in the price of the licence.

(*b*) The British Wireless for the Blind Fund will supply a free wireless set to certified blind patients, but responsibility for the subsequent maintenance of the set is the responsibility of the patient.

4. POSTAL FACILITIES. Reduced postal rates for books and apparatus.

5. BRAILLE BOOKS AND MOON BOOKS. Two varieties of printed symbols read by touch.

6. SPECIAL TOYS FOR CHILDREN.

7. TALKING BOOKS. A library service of tape-recorded books is available for those unable to read. The cost is normally recoverable from local authorities, and the machine is returned when no longer needed.

8. GUIDE DOGS. In certain cases the training of the blind person in the use of a guide dog will be considered.

The use of these dogs has a very limited application and no case can be considered unless the dog is considered to be essential to the blind person's employment.

9. VOTING. Special facilities are available for the blind to vote at an election, and they may register their votes as (1) absent voters, (2) illiterate (in which case the Presiding Officer marks the voting paper), or (3) with a companion, who is allowed to go with the voter into the polling booth.

Glossary

Accommodation: The change in focus of the eye for clear viewing of near objects. Gradually diminishes with age with the development of presbyopia (q.v.).

Amblyopia: Visual defect in the absence of apparent disease (e.g. amblyopia of disuse, in squint).

Anterior Chamber: Between cornea, in front, and iris and lens, behind.

Astigmatism: Irregular curvature of front surface of cornea, producing an aspherical surface. Can only be corrected by cylindrical lens or contact lens.

Blepharitis: Inflammation of the lid margins.

Cataract: Any opacity in the lens of the eye.

Chalazion: Meibomian cyst.

Chemosis: Oedema of the conjunctiva.

Choroid: The vascular coat of the eye, at the back.

Concomitant: Non-paralytic.

Cover test: The most important diagnostic test for squint.

Dacryocystitis: Inflammation of the tear sac.

Diplopia: Double vision.

Ectropion: Eversion of eyelid.

Emmetropia: The normal refractive state.

Entropion: Inversion of eyelid.

Epiphora: Watering of the eye.

Glaucoma: A collection of diseases characterised by elevation of the intraocular pressure.

Hemianopia: Loss of one half of the field of vision.

Heterophoria: Imbalance between the muscles of the two eyes, not amounting to actual squint.

Hordeolum: Stye.

Hypermetropia: Long-sight.

Hyphaema: Blood in the anterior chamber of the eye.

Keratoconus: Conical cornea.

'Lazy Eye': Amblyopia following squint.

Limbus: Junction between cornea and sclera.

Miosis: Contraction of the pupil.

Mydriasis: Dilatation of the pupil.

Myopia: Short-sight.

Occlusion: Covering of an eye, usually to force the use of the other.

Ophthalmia Neonatorum: Purulent conjunctivitis in infancy.

Orthophoria: Normal balance between the muscles of the two eyes.

Pinguecula: Conjunctival degenerative change at the limbus.

Proptosis: Exophthalmos.

Pterygium: A progressive opacity involving the cornea from the limbus.

Ptosis: Drooping of the upper lid.

Refraction: The state of focus of the eye. Also the estimation thereof.

Retinoblastoma: Malignant retinal tumour of childhood.

Retinopathy: Changes in the retina, usually reflecting systemic disease (e.g. hypertensive retinopathy).

Strabismus: Squint.

Sjögren's Syndrome: A group of features associated with rheumatoid arthritis.

Sympathetic Ophthalmia: Inflammatory changes affecting the second eye after a penetrating injury to the first.

Synechiae: Iris adhesions commonly following inflammation.

Trichiasis: Inturned eyelashes.

Uveitis: Inflammation of the iris, ciliary body, or choroid.

Xanthelasma: Creamy deposits in the skin; often of the lids.

Index